SOUTHERN TIMES

Contents

Introduction	
Staines to Wokingham and Beyond, Part 1, Jeremy Clarke	5
From the Archives: illustrations collated by Roger Simmonds	14
So just how many Bulleid engines were built…?	15
Guildford, 8 November 1952, with thanks to Norman Hamshere	16
London, Brighton & South Coast Railway Motor Trains	21
John Davenport. A lifelong railway interest:	
Part 2 British Railways days	25
1923 – a century on	33
Did you get it right?	
The answers to our little cartoon quiz in Issue 4	34
Stephen Townroe's colour selection:	35
Adams and Drummond Part 2	
Two Level Crossings	47
Holland Park Halt	49
William Adams '135' class 4-4-0. Nos 135 to 146	51
Southern Region people	57
Guildford (and its environs) Part 2 Colin Martin	60
Colin Tuttle – fourth generation railwayman	62
Nine Elms coaling plant – from the	
Southern Railway Magazine, February 1924	63
Electrification to Portsmouth 1937; the minutia behind the scene	64
Variety (or the lack of it) in 1966. Illustrations by Tony Harris	70
Treasures form the Bluebell Railway Museum	74
Just for amusement…	77
From the Footplate	78

The Transport Treasury

TIMES SERIES

SOUTHERN TIMES

Front Cover: Electro-diesel No E6004 brand new at Eastleigh in 1962. This was one of the first batch of six all built and successfully in service in the same year. Developing 1,600hp from the third rail and 650hp from the inboard diesel engine they were a success from the outset and for many years the only dual-mode locomotive operating in the UK. Renumbered as 73004 in January 1972, the engine survived in service until early 2002.

Above: We were recently able to access some of the black and white images taken by Gerald Daniels – whose name will be familiar to all who recall the wonderful open days at Woking, Basingstoke and Winchfield and the even better Salisbury steam events of years past. As an example of Gerald's work in earlier days we see 4Sub No 4555 arrived with a race day special at Kempton Park; the motorman seemingly about to change to the 'U' designation. Unit 4555 had originally been of three cars but was augmented with an additional steel-bodied trailer in the period 1946/7. *Gerald Daniels / Transport Treasury*

Rear cover: The all too familiar face of the rural branch line in its final days; little traffic goods or passenger and operating costs stacked against a line's retention. Here at Cranbrook C No 31588 is certainly costing more to operate than is being gained in revenue. *Gerald Daniels / Transport Treasury*

Copies of many of the images within SOUTHERN TIMES are available for purchase / download.

In addition the Transport Treasury Archive contains tens of thousands of other UK, Irish and some European railway photographs.

© Images (unless credited otherwise) and design: The Transport Treasury 2023.

ISBN 978-1-913251-48-2

First Published in 2023 by Transport Treasury Publishing Ltd.,
16 Highworth Close, High Wycombe, HP13 7PJ

www.transporttreasury.co.uk *or for editorial issues and contributions email to*
SouthernTimes@email.com

Printed in the Malta by the Gutenberg Press.

The copyright holders hereby give notice that all rights to this work are reserved.
Aside from brief passages for the purpose of review, no part of this work may be reproduced, copied by electronic or other means, or otherwise stored in any information storage and retrieval system without written permission from the Publisher.

This includes the illustrations herein which shall remain the copyright of the respective copyright holder.

INTRODUCTION

I suspect that like me, most readers will on occasions ponder on 'might have beens', as well as likely harking back to the past (well I do) view with a rose-tinted perspective.

For this issue I would like to expand upon some of my day-dreaming and also hopefully involve you as well – not exactly telepathy, but instead through the medium of emails and letters once you have read and digested what follows.

My thoughts revolve around one question, 'What might have happened to the Southern had Nationalisation not taken place'? What follows is a very personal perspective and there are several sub-topics within the one heading, notably; the network, motive power, infrastructure, traffic.

Dealing with the network first. I would suggest the closures that subsequently took place in the 1950s would likely have gone ahead anyway. Where I think alterations might have been made would have been to avoid the wholesale slaughter of lines that took place under Beeching. The Southern management was acutely aware that it needed to please its customers - here the word does seem appropriate – and that they faced a potential backlash had the withdrawal of services gone too far. One example I might give is on the Isle of Wight where I venture to suggest, Ryde- Newport – Cowes could have remained.

The situation west of Exeter is more complex. There is the oft quoted tale that Waterloo wished they might be able to divorce themselves of these lines to the GWR, but in reality we might have expected Ilfracombe, Okehampton – Plymouth and Padstow to have been retained. Torrington perhaps not.

Motive power is perhaps easier. Realistically electrification would have been expanded and probably far faster than under BR. This subject to the Southern being able to raise sufficient funds in the City at suitably attractive terms. The term the 'spark' effect comes into play here and there is every reason that the Kent Coast, Bournemouth and Salisbury might have seen the abolition of steam far earlier than actually took place. Diesel likely to have taken the place of steam, so much so that my own guess is that steam would have been eliminated everywhere probably by 1960. Mr Bulleid would of course have retried and here we will be radical, but his potential replacement, Alfred Raworth would have been a man with little appetite for steam engines that were unreliable. Consequently rebuilding of the Pacifics would hardly have been likely to have taken place and most probably most would have gone before 1960 – displaced by diesel and electric on their former duties. Whether electric engines or electric units would have taken services on some of the longer runs is debatable, personally I would say locomotive hauled trains at least for the first few decades.

Infrastructure would similarly have been cut back, disused yards and sidings no longer of use in consequence of changing traffic patterns quickly torn Up. Station rebuilding certainly, but this would have been an ongoing task anyway especially with arrears of maintenance and renewal in consequence of the legacy of WW2.

Freight is the one area which is more difficult to judge. Road competition would have hit the Southern hard but I suspect there may have been a more determined approach to retain as much as possible even in concentrated in certain areas. I doubt mail, newspapers and parcels would have been surrendered so easily. I still find it hard to believe today that so many large areas of population have no rail borne freight facilities.

So far as passenger workings were concerned, regular interval workings would surely have been the order of the day – everywhere.

Savings on staff of course, new concentrated signalling, more modern trains and the retention of buffet facilities at least. The modern day trolley service a poor substitute on its own.

Pipe dreams perhaps, but do please let me know your thoughts as well. Finally and as ever, many thanks to Andrew Royle for photo research.

Kevin Robertson

The next issue of SOUTHERN TIMES, No 6, will be available in September 2023 Contents to include: The LBSCR 'K' class, Stephen Townroe on the Isle of Wight, the Bluebell pre-preservation, SR stations in colour, a new look a the Southern in Wartime, observations at Farnborough - and lots more!

SOUTHERN TIMES

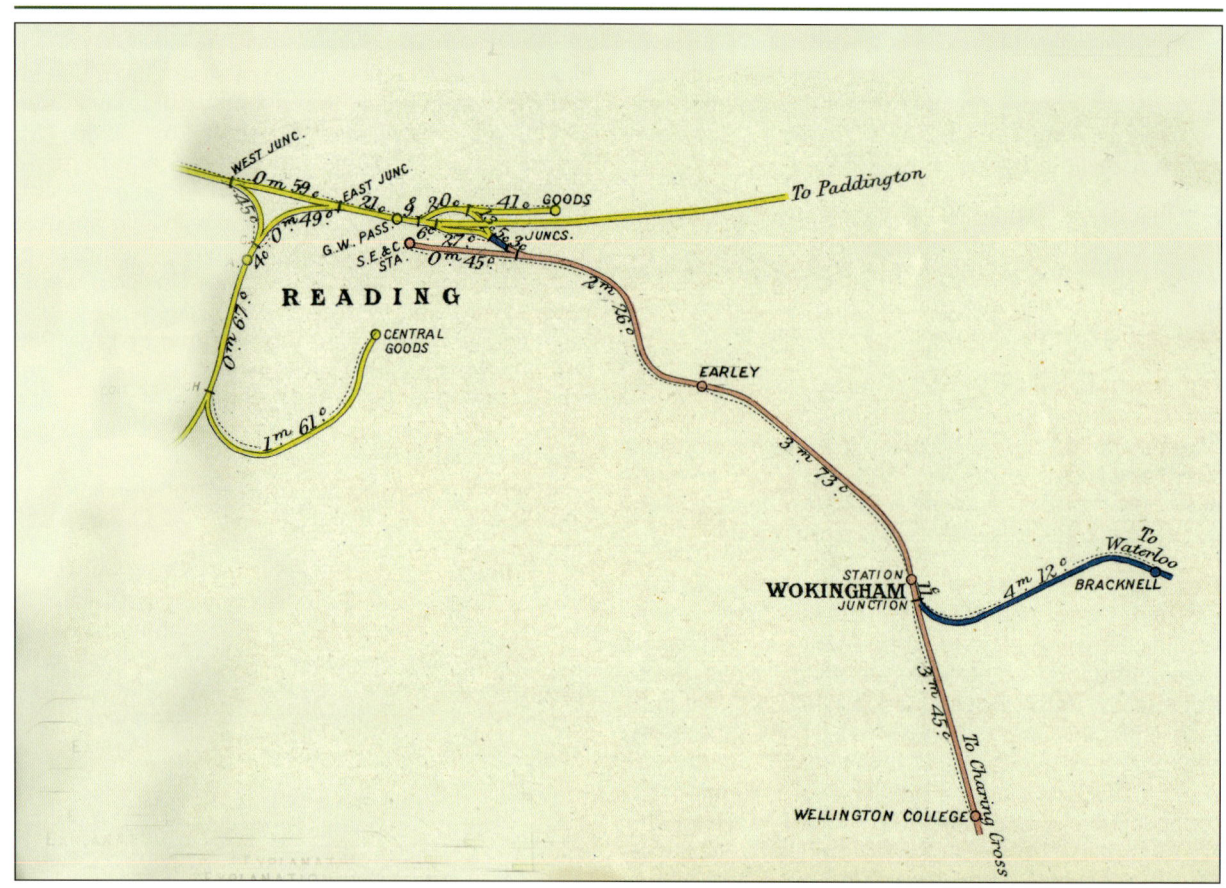

Staines to Wokingham and beyond. Part 1
Jeremy Clark

Although only eighteen miles in length the line between Staines and Wokingham had a very convoluted history for in 1845 plans were deposited for no fewer than six routes between the two towns.

The nearest of these to the one in existence today was provided by the Richmond, Staines & Newbury Junction Railway, some thirty-seven miles long. However, it would not have continued onward to serve Reading as is the practice now but continued directly from Wokingham to Newbury.
That town had been an objective for a line from Basingstoke proposed in 1844 by the London & South Western.

Though at first appearance this was a prod at the Great Western its primary importance was deemed to stem from the connection it would provide to Southampton. (The ultimate intention was to reach Swindon and, perhaps, Gloucester!) Moreover, though the Berks & Hants line, opened between Reading and Hungerford in December 1847, and would so put Newbury on the railway map the town was at that time at the end of a GWR proposed branch from the Bristol line at Pangbourne.

In the light of the South Western's plans, no less than those of the Richmond- Newbury line, it was no surprise the Great Western countered with a London, Staines, Ascot & Reading Junction Railway. From Old Oak Common this line would have got to Reading by way of Hounslow, Staines and Wokingham.

Two proposals of 1845 were of greater substance. The first of these the Reading, Guildford & Reigate Railway, that wonderfully picturesque line which runs for almost twenty miles beneath the scarp face of the North Downs. Authorised the following year, and which at its western end joined Wokingham to Reading, this had opened throughout in three stages by October 1849. It was worked by the South Eastern, which had supported the company from the first and absorbed it in 1852.

Plans for the second of these routes, a branch from Weybridge to Staines via Chertsey and Thorpe, were deposited by the LSWR. However, rather than crossing the Thames this line would have terminated on the south side close to Staines bridge carrying the road now designated the A308 over the river. The Weybridge-Chertsey section of this line opened in February 1848.

In the same 1846 session as the RG&RR was authorised, the Great Western submitted plans for another attempted incursion into the region, this time with a line from Ealing to Egham via Hounslow and Staines. Branches to the docks at Brentford and to Twickenham were also included.

The next year another independent company was authorised to build a line effectively providing the route as we now know it. This was the Windsor, Staines & South Western Railway. Though the name describes the principal objectives of the company a 'branch' from Staines to Wokingham was also included. The 'South Western' part of the title came from authority also being received for another branch from a junction on the Wokingham line at Knowle, (east of Longcross), through Chobham to a junction with the South Western main line at Brookwood. A fourth part of the Bill authorised the company to put in a triangular junction at Virginia Water (incidentally then named Trotsworth) and extend a line thence to join the LSWR branch from Weybridge at Chertsey, though the earlier powers for the direct Chertsey-Staines section were still valid at that time.

Richmond had been reached from Battersea – later Clapham Junction, a misnamed station that happens to be almost in the centre of Wandsworth! - on 27 July 1846. This line had been constructed by the independent Richmond Railway, soon to be swallowed Up by the South Western.

The WS&SWR line westwards from Richmond opened to Datchet on 2 August 1848. None of the other authorised routes were proceeded with because their construction and, indeed, other local proposals, were stymied for some time thereafter by lack of finance following the 'Mania and Crash' of the mid-1840s. However, Royal reluctance and the immediate financial problems as well as GWR opposition had all been overcome when the line from Datchet finally crossed the Thames to hug the south bank through Home Park to its pretty little terminus at the riverside in Windsor. This section opened on 1 December 1849.

Opposite top: From the 1909 RCH junction diagrams book. Key: Pink – SECR, Blue – LSWR, Yellow – GWR.

Opposite bottom: Shunting duty for 700 class 0-6-0 No 30692 at Windsor & Eton, 23 June 1957. In the background and with the circular headboard in No 32331 Beattie, which had brought the train from London Bridge via Peckham Rye, Tulse Hill, Streatham, Wimbledon, Surbiton, Weybridge, Addlestone Junction, Virginia Water and the Staines west curve to the terminus. A return to London Bridge was made via the same route. *Transport Treasury F2826*

SOUTHERN TIMES

It was not until 1851 that further moves were made in this vicinity when the Staines & Woking Railway proposed a line between these two points via Egham and Virginia Water. The next year saw the present line from Staines to Wokingham reappear as the Staines, Wokingham & Woking Junction Railway. While reaffirming most of the salient unbuilt features in the WS&SWR Bill, there was no mention of the Virginia Water-Chertsey branch. The southern part of the line from Knowle via Chobham would be made to divert slightly eastward of the original to make instead an east-facing junction at Woking, close to the divergence of the branch to Guildford. The South Western offered to work the line and to provide terminal facilities at Waterloo, subject of course to certain conditions and payments. At the same time, terms for running powers from Wokingham into the South Eastern's Reading terminus were agreed.

Authorisation of the SW&WJR routes was received in July 1853 but in November that year the company put forward expansion proposals to continue onwards from Reading to Oxford. These were prompted in part by feuding between the GWR and L&NWR for a monopoly of traffic at Oxford and, more pertinently perhaps, by the requirement imposed on the GWR that the Reading-Basingstoke branch should have the gauge mixed.

Ironically and as earlier mentioned, this line had originally been proposed by the South Western as part of a route to Newbury. It was later taken up by the previously-noted Berks & Hants Railway, soon to be absorbed by the Great Western, and opened in 1848 as a broad gauge line from Reading to an independent station at Basingstoke.

Even before this, the Gauge Commissioners had recommended to Parliament this route should feature mixed-gauge track and the order for mixing was then imposed on the GWR in 1854. Completion date was fixed at 7 February 1856 but the fine of £200 per day for failure to do this appears to have had little effect on hastening installation, for it was not until three days before Christmas that year that the Great Western finished the work.

Thus, the SW&WJR was looking to get to this route to gain the GWR line between Reading and Oxford. This was another that had had gauge mixing imposed, this time under the Shrewsbury Amalgamation Act of 1854*. The GWR main line through Reading itself remained steadfastly broad gauge only and the plan would have necessitated a short connecting link of about ½ mile in length to the west of the station: this is now Reading West Curve.

A grander underlying plan was that the Staines, Wokingham & Woking Junction, having got to Oxford, make an end-on connection there with the newly-opened – and 'narrow' gauge! - Oxford, Worcester & Wolverhampton Railway. This company had been authorised in 1845 and, after some difficult financial and political vicissitudes, had opened its line in stages by 1853. Perhaps it was as well the SW&WJR's ambitious plan failed to materialise for the OW&WR soon gained a dreadful reputation, earning itself the nickname 'Old Worse and Worse'. On 23 August 1858 at Round Oak, south of Dudley, it staged what the Inspecting Officer, Captain Tyler, described as 'decidedly the worst railway accident that has ever occurred in this country'. There were fourteen fatalities and fifty passengers seriously injured. **

Following some disagreements between the SW&WJR Board and its engineer, the notable John Hawkshaw was appointed to the post in July 1855. While accepting the basic route as a whole, he made some slight adjustments to ease the curves at Staines and Wokingham. In the former case this caused ructions among the local people who protested about the revised siting of the eastern abutment of the railway bridge over the Thames. Their concerns were overruled but to this day the B376 Laleham Road in Staines still narrows appreciably where it passes under the line.

Inspection of the completed section of route between Staines and Ascot was carried out on behalf of the Board of Trade by Colonel Wynn on 31 May 1856. Public services began on 4 June. The remainder to Wokingham was inspected three weeks later and opened on 9 July. The Reading public's immediate response to the route was lukewarm. It could not compete on either journey time or cost with the GWR's more direct line into Paddington nor, most surprisingly, was it cheaper than the South Eastern's fare for the sixty-seven mile hike to London Bridge via Guildford and Redhill. A price war followed which displeased the shareholders if delighting the travelling public. But in June 1858 a Pooling Agreement for both passenger and freight traffic was signed between the three companies.

In November that year the GWR intimated it would raise no objection to the SW&WJR making a connection at Tilehurst – no station here till 1882 - to the mixed gauge line from Basingstoke. It did, however, object to a similar proposal by the South Eastern which would have involved a westward extension from its own Reading station to Oxford Road Junction, at the south end of Reading West Curve. In the event Parliament refused the latter scheme but sanctioned the former to the SW&WJR

No 30915 *Brighton,* with multiple jet blastpipe and consequent wider chimney, is seen in deplorable external condition at Wokingham in charge of an inter-regional Wolverhampton to Dover and Eastbourne service, 28 August 1962. At the time based at Redhill, the engine will have taken over the train at Reading. It would survive just a few more months, being a casualty of the 'great cull' at the end of the year. *Transport Treasury PP1318*

on condition the GWR did not itself start to build such a line within a year of authorisation. If it should do so, the Staines company need only complete the link to SER metals east of Reading.

Not surprisingly the Great Western preferred to do the work and laid a single standard gauge track passing north of Reading station on a gradient falling to the east. Having borne away from the line to the lower goods yard, it tunnelled beneath the GWR main line to join the SER at Reading Junction, a half-mile from that company's terminus. The Great Western maintained ownership of this line throughout except for the five chains immediately south of the tunnel that formed the authorised SW&SJR link to the South Eastern. The opening for goods traffic on 1 December 1858 had been delayed by three months following the decision to install double track instead. Passenger trains began to travel the line from 17 January 1859. The South Eastern later gained permission to use the link on payment of tolls.

Another connection actually into the GWR station opened on 17 December 1899, an incline being constructed on the south side of the main line from a new 'Low-Level Junction' at the northern end of the former SW&WJR five chain-long link. (An additional incline was opened in June 1941 as a wartime measure. This was slightly further east still, the object being to facilitate the movement of freight traffic between, particularly, the SR Feltham Yard and the GWR down side Scours Lane and up side New yards.)

The close association between the Staines, Wokingham & Woking Junction and the London & South Western had become more formal in July 1858 when the LSWR took a 42½-year lease on the local company. In the event this situation lasted for only twenty years, the former absorbing the latter in 1878. Meanwhile, with the blessing of the SW&WJR directors, Hawkshaw had drawn up the plans for the Egham & Woking Railway route heading south from the junction at Knowle. These were deposited in November 1859. However, nothing further was done with this line though the SW&WJR Board decided neither to support nor oppose the promotion of independent Bills for two routes starting on the GWR at West Drayton, one of which affected it directly.

The first of these lines would have probably followed a route similar to the later Great Western branch down the Colne valley to Staines via Colnbrook opened over twenty years later. But from

SOUTHERN TIMES

Railtour duty on 18 September 1960. The LCGB South Western Limited but with a certain amount of licence as it covered some non-LSWR lines and involved former LSWR, SECR and S&D locos. L class No 31768 is seen here amongst the neat layout of Ascot waiting to set off towards Aldershot, Alton, Winchester and Eastleigh. *Transport Treasury*

Staines it proceeded onward to Chertsey on much the same line as the original but un-built LSWR branch authorised to terminate south of the river. There would appear to have been no connection made at Staines. By contrast the second line proposed to use the SW&WJR from a point some ½ mile east of Staines almost all the way to Knowle to take up the powers of the line to Woking.

The first proposal was rejected for failure to meet Standing Orders while the second found favour with the Staines Board but not that of the South Western. The section north of Staines having been withdrawn, the Commons passed the Bill for the southern part of the route but no work was actually done and the powers lapsed.

The year 1863 saw further flurries of proposed and planned activity in the area. Much to the mystification of the Staines, Wokingham & Woking Junction directors, the Berkshire Chronicle carried notices in November that year referring to two more local independent companies setting out their stalls. These were the Sunningdale & Cambridge Town Railway and the Sunningdale & York Town Railway, both of which would have ended in parts of present day Camberley.

Not to be outdone, the Surrey Gazette intimated the LSWR was about to construct the remainder of the Weybridge-Chertsey branch but to Virginia Water rather than Staines. As the South Western made no response to Staines company enquiries about this proposal, the SW&WJR Board decided to oppose it if only to get a Parliamentary hearing. At the same time the directors registered opposition to yet another Staines, Egham & Woking Junction Railway. In the event they later agreed to the Chertsey branch on the understanding that 'traffic passing through Virginia Water would pay mileage as though it has passed via Staines'. A similar caveat attended the Staines Board's agreement to the Sunningdale & Cambridge Town line when the Chairman became a director of that company. The South Western also withdrew opposition before the Commons passed the Bill in April 1864. Among the clauses in the Staines, Egham & Woking Junction Bill was one that stated the company would widen the SW&WJR line to four tracks between the junctions at Staines and Knowle. Great ambition, faint hope!

The South Western's extension from Chertsey to Virginia Water was sanctioned in 1864 and opened on 1 October 1866. (The temporary station at Virginia Water, opened with the line, had by this time been superseded by the permanent one on the same site.) A new station was provided at Chertsey, on the west side of the Guildford Road. In the same Parliamentary session, an extension of the Sunningdale & Cambridge Town line to Aldershot was put forward though it was some time before this was achieved.

Yet another attempt to provide a north/south line in the vicinity was proposed in November 1870 by the Windsor, Ascot & Aldershot Railway. This line would have crossed the SW&WJR at Ascot. The South Western was vigorously opposed to this plan and, together with the Staines company, successfully petitioned against it. However, as a result of this proposal there arose the question of the convenience of parts of the LSWR timetable in the Aldershot area. Such was the dissatisfaction registered by the travelling public, the company resolved 'that in order to prevent the promotion of any more competitive schemes of railway in opposition to the Staines, Wokingham & Woking Junction, the General Manager of the LSWR be requested to give such further passenger trains as shall meet the wishes of the Windsor, Ascot & Aldershot Railway Company'.

Whatever the direct outcome of that resolution, we find the South Western obliged to revive parts of both the WA&AR and Sunningdale & Cambridge Town Railway schemes in 1873. These moves appeared as the line from Ascot to Ash Vale. The first part of this provided eastbound access to the LSWR main line east of Farnborough and opened on 18 March 1878. (The single line from what became Frimley Junction to Ash Vale Junction and access thence to Aldershot did not open for another fifteen years. The trackbed of the Ash Vale link was, in fact, built for double track though a second line has never been laid. The east and west connections to the main line south of Frimley Junction were both taken out of use on 24 October 1964.)

Such moves clearly failed to satisfy all for three more schemes in the area arose in 1881 and 1883. The Windsor, Aldershot & Portsmouth Railway would have crossed the SW&WJR at Bracknell from where a branch was planned to Ascot. That station would have featured as the crossing point of the 1883 proposal, the Windsor & Aldershot Railway. Yet another 1883 promotion, the Staines, Chertsey & Woking Railway, would have intensified the rivalry between the LSWR and the GWR. The company intended to have an independent terminus at Staines but connect with the Great Western's branch then under construction from

'Normal service is resumed', N class No 31821 in charge of a down (Reading bound) freight at Wokingham on 1 June 1963. The bracket signal at the end of the opposite platform has the arm clearly for a train to take the line towards Ascot. The arm to the right was for trains over the SECR line to Farnborough and on to Guildford. *Transport Treasury F1787*

SOUTHERN TIMES

West Drayton. Thereafter it would have struck southward via Thorpe and Chertsey to join the South Western line just to the east of Woking. It also sought running powers on to Guildford. None of these got further than basic proposals, the system as it is today, bounded by the Staines/Reading/Aldershot triangle, having almost been completed.

Electrification came to all the lines considered between the two World Wars: from Whitton and Hounslow junctions to Windsor on 6 July 1930, Hampton Court Junction to Chertsey and Staines on 3 January 1937 and Virginia Water to Reading South, Ascot to Ash Vale, Frimley Junction to Sturt Lane Junction and Aldershot to Guildford, all on 1 January 1939.

Though nowadays the inclination is to consider the Staines-Wokingham line the 'main' one, it is significant the section leading to it from Waterloo is still known as the Windsor Line. The fact the originating company boldly put Staines and Windsor at the beginning of its title suggests that, before considering the line to Wokingham in any detail, it might be politic to take a trip to Windsor first.

The station at Staines opened with the line on 22 August 1848. It had the usual side platforms and a very handsome two-storey yellow-brick building on the up side, still in use. The down side had little more than a brick shelter until the Southern provided a contrasting single-storey red brick structure with booking office, waiting rooms and staff quarters, all shielded by a typical SR steel canopy. There was a small two road goods yard on the up side immediately at the London end accessible from the forecourt; it had side- and end-loading facilities as well as a weighing machine and a large timber goods shed. This survived until 1988 despite the sidings having been taken out of use fifteen years earlier. The sidings of East Yard lay beyond the Kingston Road bridge, also on the up side, used mainly for sorting. Four of them were relaid and electrified in 1974 for stock berthing although two such sidings already existed at Chertsey on the site of the loco shed closed in 1937. The reason behind the move, which permitted Chertsey sidings to be lifted and the land eventually sold, was that it lay at the extremity of the area to be controlled by the new Feltham signalling centre. It was deemed wiser to move the sidings fully within the area; Staines happened to provide the preferred location.

On 9 August 1957 a loco movement from the East yard to the down line resulted in a collision at a converging speed of about 30mph at the crossover to the east of the Kingston Road overbridge. The 12.02pm Weybridge/12.07pm Windsor-Waterloo service, two 4EPB sets combined at Staines, had started away against signals. The engine, '700' class 0-6-0 No 30688, came off worst, finishing up on its side across the down line and the adjacent loop siding. Casualties fortunately were light, only the driver of the engine being detained in hospital with a broken leg. His locomotive, however, was not so lucky, being written off due to the severity of the damage. The driver of the electric train was held primarily to blame, though the actions or inaction of some station staff were criticised.

The station had been served by several signal boxes, growing to five in number as the complications of the layout increased, until two electro-pneumatic boxes superseded them in 1904. East box stood on the down side opposite the East sidings while West box was located in the triangle of lines to the west of the station. Both of these worked until 1930 when a new box of undoubted LSWR parentage was provided by the Southern alongside East box. This worked until Feltham took over its control area in September 1974.

At the junction immediately beyond the station's down end, at 19 miles and 6 chains from Waterloo via Richmond, the Wokingham line bears away on a sharp curve with a permanent 20mph speed limit. There were, for many years, two berthing sidings on the up side of the Windsor line, later electrified. A further quarter mile on brought the junction with Staines West Curve which came into use in April 1877. Seven years later Staines High Street station was opened immediately beyond it and to the west of the bridge over the town's High Street. Clearly complaints had been received that any train using the curve did not provide an opportunity for passengers to board or alight at Staines. The station was closed in February 1916 as a wartime economy measure and never reopened; regular use of the curve by passenger services did not survive either. However, the station was not demolished until 1932 and the curve itself did not finally succumb until 18 March 1965. Much of its route has since been taken by South Street, a ring road that carries traffic away from the now-pedestrianised High Street.

#Staines station itself had the suffix 'Old' added in 1885 to differentiate it from High Street but that was changed to 'Junction' in 1889. In 1923 the new Southern Railway changed it again, to 'Central', even though High Street had been closed for almost seven years and never reopened. There is no suffix now, it having been suppressed in 1966.

The Great Western branch from West Drayton opened to Staines in 1885 though it was refused entry to the South Western station. Instead it

On the same day and at the same location, No 33023 was recorded with a Guildford to Reading South passenger working.
Transport Treasury F7817

terminated close to the northern end of the road bridge over the Thames, the station building being a converted Georgian house to the south of the Windsor line. There was no direct connection between the two though both had access to the Staines Linoleum Company's factory opened in 1864. However, in 1940 a link was constructed from south of the GWR's Yeoveney Halt to provide an alternative route via Staines for north/south trains should the main such routes through London become impassable. It was taken out of use in 1947 but remained in situ until 1959.

The Western Region withdrew branch passenger services in March 1965 though freight continued to run, particularly to an oil terminal that had been established in Staines West goods yard in the early 1960s. But following the complete closure of the branch south of Colnbrook in 1981, oil traffic arrived at the Southern's Staines station and shunted into the terminal through a new link north of it. This traffic continued until 1991 when the terminal closed. Much of the track, including the connection, remains in place though long unused.

Three-quarters of a mile beyond Staines the line crosses the Wyrardisbury River, which also marks the point where the GWR branch turned towards the north and bridged the Windsor route. Three quarters of a mile further still, the M25 bridges the line and at 21½ miles the track crosses the Colne Brook and comes to Wraysbury. This station opened with the line though it was moved from the north to the south side of the Coppermill Road overbridge in 1860. The very typical South Western brick station building was on the up side, its place now having been taken by a nondescript flat-roofed brick building with ticketing facilities in the open alongside it. The goods yard, on the up side at the London end of the station, consisted of three sidings with a goods shed, weighing machine and a crane. The signal box was located right by the points where the sidings merged into the up line. It closed in mid-1964, traffic having been withdrawn from the yard two years beforehand.

The station stood and, indeed, still stands on the eastern fringe of the village though there have been subsequent ribbon developments along Coppermill Road towards Horton in the shadow of the massive 1970-opened Wraysbury Reservoir. (It has a capacity of 34,000,000,000 litres / 7,490,000,000 gallons.) A mill had first been established on the Colne Brook upstream of the site of the railway bridge in Saxon times. It began to be used for paper manufacture early in the 17th century though from about 1720 it was at various times used for milling iron and copper. By the mid-19th century, almost coinciding with the opening of the railway, paper was being produced once more, much of it of very good quality. It is probable transporting goods by water from the wharf in the mill's vicinity was superseded by movement by rail, perhaps accounting for weighing and craning facilities in the goods yard, unusual equipment for a small and rather remote village station.

There are frequent open stretches of water, usually flooded gravel pits, as the line continues on between Wraysbury and Sunnymeads (22m 48ch). This station is situated just to the north of the bridge carrying the B376 Staines-Slough road over the line. It is provided with a single island platform and was opened by the Southern Railway on 10 July 1927. Steps and a footpath from the bridge lead to the former ticket office building on the down side. This has been out of use for many years – the station has been un-manned since 1946 – ticketing facilities now being provided by an up-to-date machine beside it. A footbridge over the down line leads to the platform. As a point of interest Sunnymeads is among those stations in the south-east of England with a regular train service seeing the fewest number of passengers.

A loop of the river is almost under the track soon after leaving Sunnymeads but it bends away south westwards before Datchet, 23½ miles from Waterloo and the temporary terminus until the line onward to Windsor was opened. The station is in the centre of the village, its two-storey station master's house on the up side being reminiscent of that at Staines. This survived the destruction by fire of the original adjoining single-storey station building in 1986, that having since been rebuilt in a simpler style with residential flats above it. The down side featured the smallest of timber shelters which has been superseded by one of steel and glass with a ridge roof. A similar structure fronts the station building. Goods facilities were provided on both sides of the line at the up end. The up yard consisted of two sidings with a goods shed and coal yard as well as a side- and end-loading bank. The down side yard was also equipped with two sidings. Unusually, headshunts were provided for both of these yards. Goods facilities were withdrawn in January 1965 though the goods shed survived for another thirty years.

The signal box was at the end of the down platform beside the level crossing over the High Street. Only five chains further on the line passed over Queen's Road on Mays Crossing, where the gate box was also on the down side but to the west. The gates at High Street were superseded by lifting barriers in 1973 and on 17 December 1974 control of both crossings passed to the Feltham control centre.

The line to this point has been virtually dead straight towards the north-west since the junction with the west curve at Staines, and continues to be so for another half-mile. A slight kink then gears it up for a long anticlockwise sweep through some 150° until ultimately settling to face south-west at the Windsor & Eton Riverside terminus 25½ miles from Waterloo. In the course of this curve a three-span girder bridge – Black Potts Viaduct – of eleven chains length in total from mp 24¾, takes the line to the south side of the Thames and into Home Park.

The station is built on a curve matching the beginnings of a loop in the river as it turns from facing south to north-west. The 'Riverside' suffix describes its position perfectly, for the northern extremity of the goods yard was right on the bank and the station itself sits in the shadow of the castle. (I was once asked in all seriousness by an American lady why the castle had been built right under the Heathrow flight path. True!) The quite magnificent station building alongside the southernmost of the three platforms was designed by the celebrated architect William Tite. It is in red brick with stone arched doorways and mullioned windows. Much of the stone is carved, including the chimneys. The opulent Royal waiting room is at the eastern end. The many doors on to the platform through the curtain wall alongside Datchet Road are both wider and higher than normal to accommodate the mounted pageantry that went with many monarchical movements.

Three platforms were provided, two of them under an overall roof though this did not stretch their full length. A very spacious circulating area still exists between those platforms and the splendid booking hall though that is now a wine bar. The northern of the two platforms was an island whose outer face formed the third platform facing the goods yard, the overhanging eaves of the roof giving it cover. Until electrification reached Windsor in July 1930 the two roads under the train shed had scissors crossovers between them for engine release. The now-lifted outer platform road was also provided with this facility though in this case it lasted until 1965 to cater for loco-hauled excursion traffic. Ten years were to elapse before a release crossover was laid between the covered platform lines.

The goods yard and loco depot were located between the station and the river, though a side- and end-loading dock was provided at the outer end of the down side platform, presumably for royal horses and carriages. There were five sidings in the goods yard, one of which passed through the substantial brick goods shed while another had a weighbridge in it and a third, next to the loco shed, was flanked by a side-loading bank. A crane of 9 tons capacity was also provided. Goods traffic was withdrawn in April 1965.

The loco depot was north of the yard and consisted of a brick-built two-road shed under a pitched and slated roof with the offices and stores topped by a large water tank adjoining the south-east corner. An additional siding backing on to coal pens for local merchants was interposed between the shed and the turntable road with its coal stage. In the late 19th century this 50' turntable was installed further north than the original located outside the shed, which accounts for the unusual position of the coal pens in relation to it. The South Western allocated between five and eight engines here until electrification though the shed was still working in the later 1930s. It did not survive the war however, being out of use by 1942. The whole area once occupied by the yard and loco depot is now a car park.

The signal box was sited on the up side by the entrance to the goods yard and worked the adjacent level crossing that gave access to the riverside. Both box and crossing were abolished in September 1974 although a footbridge was provided for pedestrian use. Vehicular traffic is now routed via a new road constructed along the northern fringe of the car park.

Following electrification it was common for the branch off-peak service to be part of combined Waterloo-Weybridge/Windsor trains divided or joined at Staines. These were for years worked by the venerable 2NOL sets until 2EPB stock took over in the late-1950s. Franchised to South West Trains, the service to both Windsor and Weybridge is now half-hourly throughout the day, Monday-Saturday. Windsor trains call at all stations west of Twickenham inclusive and Richmond, Putney, Clapham Junction and Vauxhall intermediately to the east thereof. Sundays see the same service level but calls at Wraysbury and Sunnymeads are only hourly. Weybridge services now run via Hounslow calling at all stations. (On Sundays this service is hourly.) Thus, together with Reading line trains calling, Staines sees six workings each way in a weekday off-peak hour.

To travel the line to Wokingham we need to return to Staines. As already noted the station building here dates from the line's opening with subsequent SR additions. A refurbishment in 2008 was followed from August 2011 by a much more comprehensive updating.

For twenty years and more after opening, the station continued to stand in quite rural surroundings and it was only in the later Victorian era that housing development began in the vicinity. Gresham Road commemorates the farm that bordered the line to the south until the latter part of the 19th century when property began to be built on it. The West Yard bordered the road also, the bend in its sidings tightening as the Wokingham line climbed and curved south-westward toward the Thames bridge. This curve is reversed for an equal distance on approach, the river being crossed on a three-span seven chains-long girder structure supported on cast-iron columns. The yard was later expanded to provide private sidings for a company of steel fabricators who still use it for storage although all goods traffic was withdrawn from the station in August 1971.

Having reached the south bank of the river the line, still curving but now to the west, falls sharply to pass over Thorpe Lane crossing at 19 miles and 67 chains from Waterloo. The crossing box was on the down side west of the road and was also a block post, a function withdrawn after the Feltham signalling centre came on-stream in September 1974. Barriers had replaced the gates fourteen months earlier and these were worked by Feltham from March 1975. The next crossing, Egham Causeway, was unusual in that the gates were normally set against road traffic. The box was on the up side beyond the road but was put out of use when the crossing was closed completely on 29 September 1970. Pooley Green crossing, now within the shadow of the M25 bridging the B388 Egham-Thorpe road, also had its box on the up side. It lasted rather longer than the preceding pair, not being closed until 8 May 1977 following installation of barriers in place of the gates.

To be continued in Issue No 6.

From the Archive;
Illustrations collated by Roger Simmonds

We are grateful to our long term friend and contributor Roger Simmonds for submitting four more images going back – well a few years.. ..

The crossing keeper's cottage at Crampmoor Crossing between Chandlers Ford and Romsey on the line from Eastleigh. This view dates from c1880 and shows the then crossing keeper Fanny Noyce.

A cautious approach through flood water at Walkford between New Milton and Christchurch, 12 December 1907. The engine is an Adams 0-4-2. (Contemporary records indicate the rainfall for the autumn of 1907 had been considered exceptional.)

A rare beast, the station at Gosport Road on the short branch to Stokes Bay which closed in WW1. One day perhaps a quality image of a train on the line to Stokes Bay may materialise.. . .

So, just how many Bulleid engines were built?

Students of the Southern Railway will immediately be able to answer the above question (some perhaps naming each). Similarly, it is not phrased as a deliberate trick hence we are not talking of an original and a rebuild counting as two. Hence to set the ground rules let us confirm:

30 x Merchant Navy
110 x Light Pacifics
40 x Q1
5 x Leader (generalising here as although only one steamed, two were part complete, and the material was present for the remaining two)
2 x Electric locos
3 x Diesel / Electric
Not forgetting of course the Turf Burner in Ireland.

As mentioned at the start, we do not include rebuilds and modifications to other classes.

By far the largest type numerically then were the Light Pacifics, 'West Country' and 'Battle of Britain' types with minor modifications as the build progressed over a six year period. 110 engines in total erected primarily at Brighton and with some at Eastleigh. So what if we were to now say, that figure of 110 is in fact wrong as we now have proof of No 111 erected as it turns out at Ashford.

However, before you start rushing to your duty Ian Allan ABC from the 1950s let me also say, what was built at Ashford was in outline at least a Light Pacific but not perhaps otherwise to the accepted form.

The story comes from enthusiast Norman Hamshere who informs us that in 1951 the apprentices at Ashford were tasked with the building of a 2' gauge outline model of a Bulleid Pacific to display in the town. This was duly completed although it should be said as it was an outline model the locomotive and tender shared the same fixed frame; hence we were cautious at the start of this piece not to say ' how many Bulleid Pacifics' had been built. As to why it was given the number 34055 is not known, but after it had served its purpose it was evidently stored before being eventually purchased by Mr Hamshere (who is also a very capable engineer) and who converted it from a static model to one operatedby battery power where it now operates on his garden railway.

Above: The 141st Bulleid Pacific? (Well perhaps not quite.) No 34055 (2) on display in Ashford (Kent) High Street in 1951. Remember the Westminster Bank and shop blinds as seen….?

Left: After years of neglect, the second No 34055 as restored by Norman Hamshere. Previously built as a static display, the engine now operates with an electric motor. *Norman Hamshere*

Guildford 8 November 1952
With thanks to Norman Hamshere

The accident that occurred at Guildford on 8 November 1952 was a tragedy in more ways than one. One railway employee, the motorman of the EMU involved and a passenger on the same train died whilst 17 others were injured or suffered from the effects of shock.

The circumstances at Guildford itself are easily told; the EMU set running away out of control on the falling gradient from the direction of Aldershot and colliding with the tender of a 700 class No 30693 which was legitimately crossing the path of the electric set – the latter having run past two stop signals at 'danger', the driver remaining at the controls but powerless to slow the speed of his runaway train.

In general terms accidents occur due for one of three reasons;

Human error
Equipment failure
Outside interference.

In the case of Guildford it was the second of these reasons and one which tragically could not have been foreseen – although British Railways were quick to respond as a result. In short it was also an accident waiting to happen and purely by chance a similar incident had not occurred previously.

The vehicles involved were the two coaches forming 2Bil set No 2133 built in 1938 with steel underframes and bodies made of composite construction having steel panels over a wooden framing. The Westinghouse air brake was fitted whilst it should be noted a handbrake was only provided in the respective drivers cabs at either end and not in the guard's compartment. As with all SR EMU sets. they units could also work in multiple under the control of a single driver.

On the day in question, No 2133 had previously formed the 7.28pm Reading to Waterloo service. It is relevant to mention that it operated as a single 2-car set between Reading and Ascot where a second 2-car was attached to the rear. Arrival at Waterloo was without incident at 8.42pm and where after a 12 minute lay-over, the same formation left again for Reading at 8.54pm.

The pair of 2-car sets arrived at Ascot at 9.44pm and after the front two cars had been uncoupled the front portion departed, again for Reading. Shortly after set No 2133 now in charge of Motorman Tullett set off for Guildford as the 9.46pm departure. Up to

Possibly the only photograph taken of the aftermath of the accident. The shortening in the length of the coach is sadly all too apparent. *Norman Hamshere*

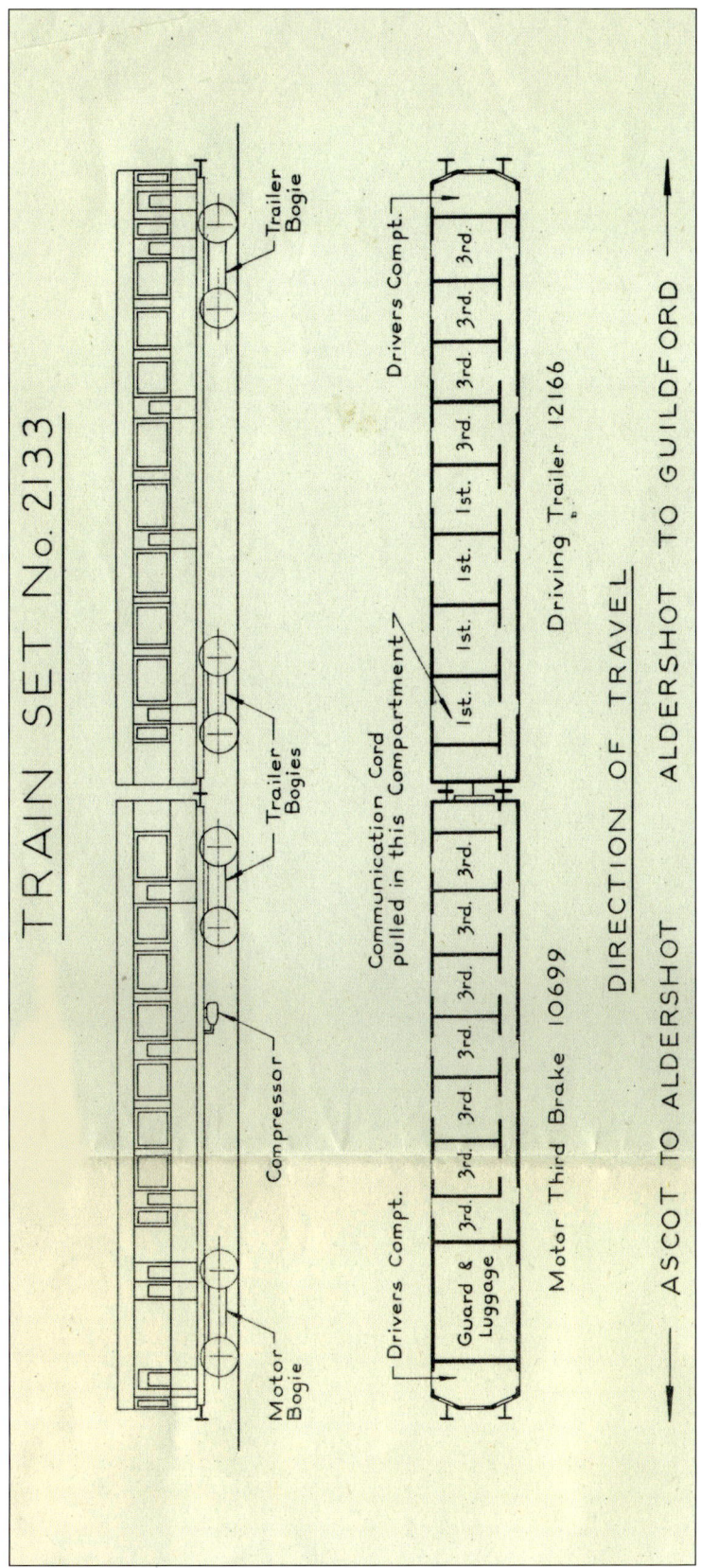

that point, none of the drivers involved had reported any issues with the brakes or any obvious drop in air pressure.

Guard Liddington who was in charge of the train, reported he was in the compartment immediately behind the driver's cab and that he tested the brake by pulling Down the emergency handle until only 25-30psi was showing on the gauge. He heard the sound of the escaping air and noted the pressure returned to the normal 70-75psi level when the handle was restored to its normal position. Confirmation of this test was subsequently made to the enquiry by the porter at Ascot.

Subsequent to leaving Ascot, all was normal with No 2133, with stops made at Bagshot, Camberley, Frimley, Ash Vale and Aldershot where again as was practice for this service, the train reversed direction. Motorman Tullett changed ends although Guard Liddington remained where he was, a fact that probably saved his life. Further stops were made at Wanborough and Pinks Hill, none of these nor the earlier speed restrictions necessary for collecting and giving Up the single line tablet at Frimley Junction and Ash Junction respectively, nor the 30mph restriction from Ash Junction to Wanborough were other than normal.

It was only upon the final descent from Pinks Hill (summit) to Guildford that there was any cause for concern. Instead of the brakes being applied at the distant signal as might be expected, there was no slackening in speed. Looking through the periscope Liddington saw the red indication of the outer home signal flash past at which point he attempted an emergency brake application but to no effect. Equally important was what he stated later that he had no recollection of hearing the air compressor working – but then he might be excused this in that it was such a normal occurrence – and also that he did not remember looking at the pressure gauge throughout the journey – even it seems at this point. The driver and guard of a train passing the opposite way to the runaway did not report hearing a whistle from the runaway.

The final consequences were as might be imagined. The runaway continued past the Guildford inner home signal also

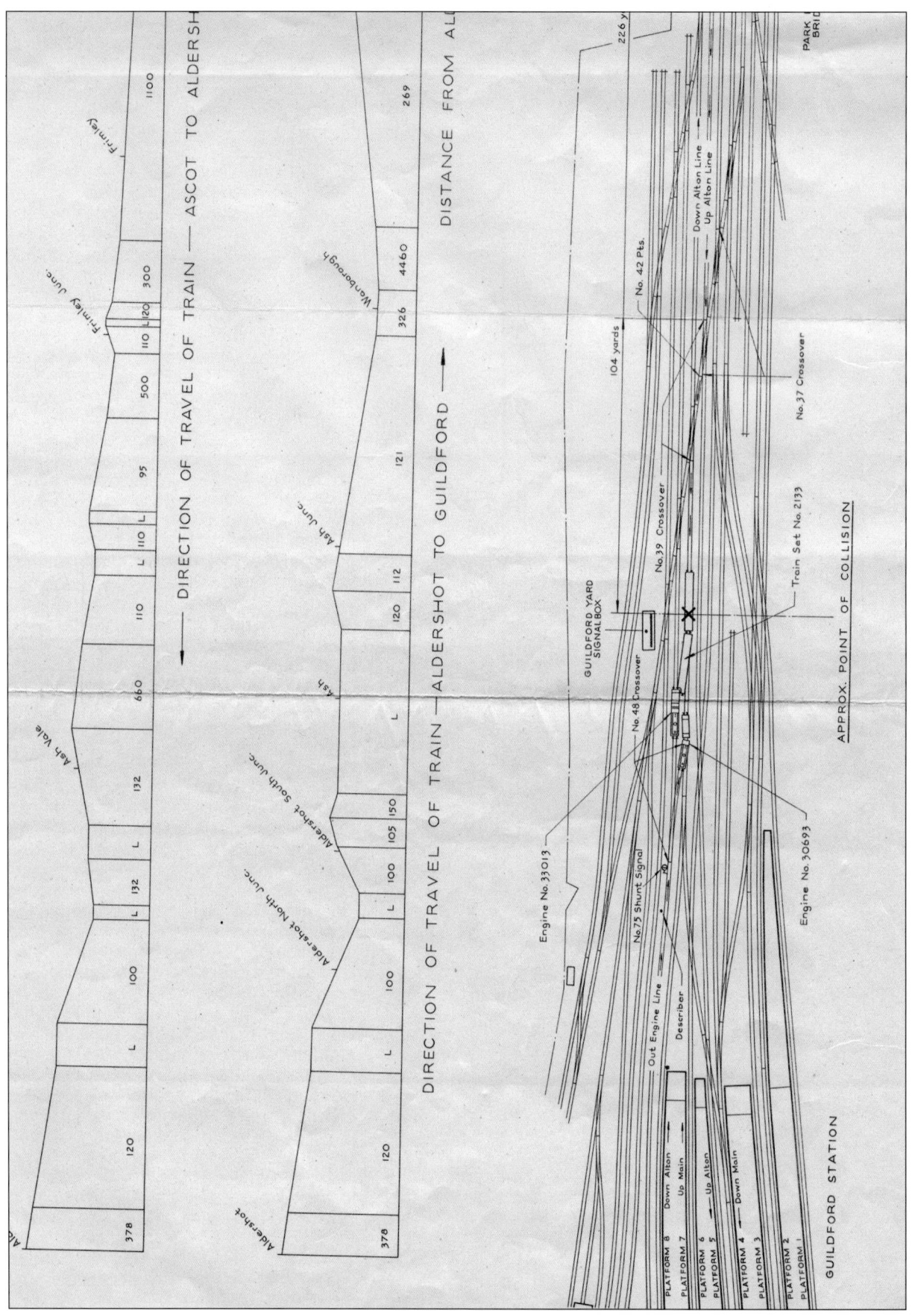

showing danger, before colliding at a calculated 55 mph with the tender of the steam engine as previously described. On board the train the enquiry noted a passenger – it was never established who – had pulled the emergency chain during the descent, but of course with no air available it had no effect.

The accident was investigated by Brigadier Langley with the cause relatively easily established. This was that the compressor fuse had blown on the set so preventing the compressor working and in consequence the air reservoir was slowly being depleted. Subsequent tests indicated that even with depleted air normal stopping would been unaffected. This loss of air would though have shown Up on the dUplex gauges in driver's cab and also on the gauge in the guard's compartment.

Brigadier Langley commented Upon Motorman Tillett's previously clear record and could only conclude that he had got so used to the effective of the Westinghouse brake that he simply failed to confirm his efficiency by checking the gauge. It must have been a terrible shock to suddenly realise he was experiencing a total brake failure but even if he had applied the handbrake – in effect little more than a parking brake (there was no evidence this has been done) the resultant reduction in speed would have been minimal.

As to why the all important fuse blew was never fully established but it was considered this could only have occurred after the train had left Ascot as a 2-car set on it final fated journey. A brief surge caused after gapping might well have been the initial catalyst as subsequent examination indicted there had not been a short circuit.

Brigadier Langley's conclusions were that something like a control governor, as indeed was already fitted to more modern stock, might well be included in existing units currently equipped with just the ordinary Westinghouse brake. This device is connected into the main train (air) pipe and is such that power is cut off to the traction motors, and similarly cannot be re-applied, should the air pressure fall below a specified limit. This was accepted by BR and all electric units not just on the Southern were suitably modified.

So far as the aftermath was concerned, the accident was reported as having occurred at 10.34pm. In addition aside from the two fatalities and injuries sustained within the electric unit, the driver and fireman of the steam engine were also slightly injured. The emergency services arrived promptly with the first ambulance taking away the most seriously injured just 13 minutes after the collision. The last of the injured conveyed to hospital just 56 minutes after the crash.

As was then the norm, recovery, repair and the restoration of normal working was the priority. Ignoring the locomotive and electric unit, damage otherwise was slight with just a single pair of points which had been run through bent, together with the controlling facing point lock. Three passenger trains had of necessity to be cancelled but arrangements were made to convey passengers (likely by road taxi) to their required destinations. The local breakdown crane based at Guildford was brought into use as soon as the injured had been removed and normal services restored for 7.00am the next day, signalling being fully operational again at 11.13am.

As we have reported before, the resilience of the railway to deal with an incident of this type yet again shows through. Today we might well expect the complete area to be closed to all traffic for a minimum of one to two weeks. That is not to criticise either the events of 70 years ago or today but simply to effect comparisons.

Having also occurred during darkness, we have no record of any official images being taken, but come daylight it was a different matter. Enter then the story one young schoolboy, Normal Hamshere, who having learned of the accident was able to record the attached image around 8.00am the following morning and from which we can see the damage to S12166 as the leading coach of the 2Bil set. Not surprisingly it was officially withdrawn just one week later on 15 November and subsequently scrapped, set No 2133 subsequently running with a post-war all steel HAL trailer.

Postscript; Your editor recalls many years ago travelling late one night as a passenger in the front coach of a REP set out of Waterloo which formed the second four-car portion of a Bournemouth line train. With just eight coaches the train was moving fast that is until passing Earlsfield when there was a bang and a shower of flames and sparks from under one of the REP set coaches. The service was almost empty and so with no one in officialdom seemingly aware, he made his way to alert the guard who had been travelling in his compartment in the front four car set. Guard returned and clearly spoke to the driver with the train making an unscheduled stop at the usually unused island platform at Walton-on-Thames. Meanwhile sparks and flames had continued to emit from under the coach. It transpired the air compressor was at fault and this was isolated after which the train continued on this way, air for the train brakes now supplied from just a single compressor under the front four coach (TC) set.

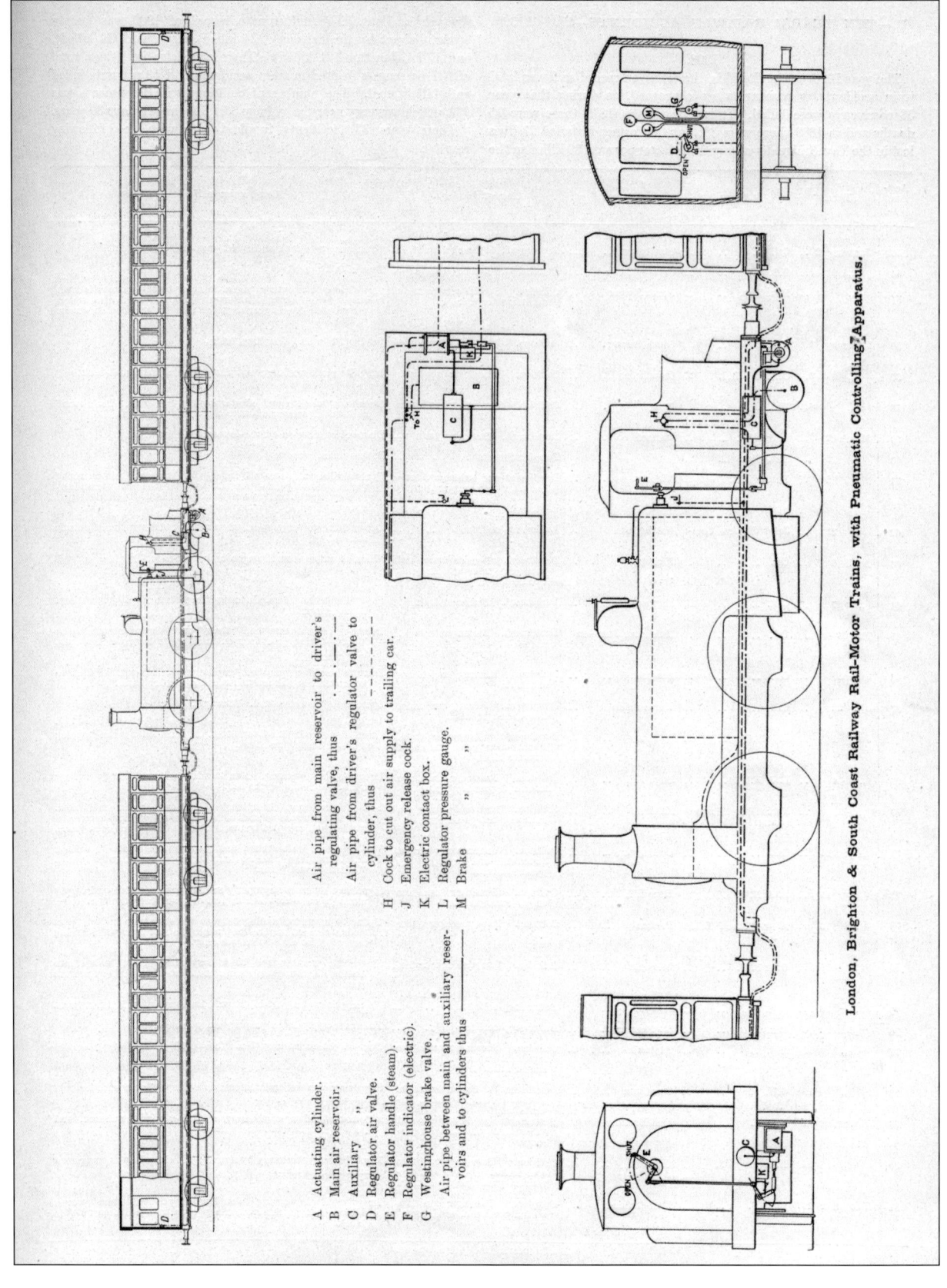

London, Brighton & South Coast Railway Rail-Motor Trains, with Pneumatic Controlling Apparatus.

London Brighton & South Coast Railway Motor Trains

Reprinted from the 'Railway Gazette' 13 November 1914

Pneumatic control of the engine operating mechanism

We are indebted to Mr L B Billinton, Locomotive Engineer of the London, Brighton & South Coast Railway, for the information, which follows, relating to the motor trains employed on that line. The principal interest in the arrangement of these trains is that the control mechanism is pneumatically operated instead of by the more usual system of rods, levers and similar mechanism. The engines employed for working the London, Brighton & South Coast Railway motor train services are side-tank engines of two classes -The 'A' class, 0-6-0 type, and the 'D1' class, 0-4-2 type, both built by the late Mr Wm. Stroudley. These engines have all been in the service for over 30 years and are admirably adapted for the class of work on which they are now employed.

The 'A' class engines have cylinders 12 in. diameter by 20 in. stroke; wheels, 4 ft. diameter; working pressure, 150 lb. per sq. and a mean weight of 24½ tons. They are used on the lighter motor services, taking a 54 ft. bogie corridor coach, seated for 60 passengers, with a guard's and driver's compartment at one end and weighing 24½ tons unloaded. Occasionally a trailing coach is also added to these trains, but the ordinary service is one coach.

The 'D' class engines have cylinders, 17 in. diameter by 24 in stroke; coUpled wheels, 5 ft. 6 in. diameter; working pressure, 150 lb. per sq. in.; and a mean weight of 38½ tons. These are used on trains composed of two or four coaches, with the engine sandwiched in the centre. The smaller trains consist of two 54-ft. bogie compartment composites, each seated for 24 first-class and 40 third-class passengers, with a compartment for the guard and driver at one end and weighing 23½ tons unloaded.

The four-coach trains consist of two 54-ft. bogie composite corridor coaches and two 54-ft. bogie corridor thirds, with a compartment for the guard and driver, the latter being at each end of the train. The total seating capacity of the train is 36 first-class and 200 third-class passengers, and the total weight of the four coaches is 93 tons.

The performances of the 'D' class engines in hauling the motor trains may be gathered from the following:

Crystal Palace and Coulsdon Service.

Maximum load (exclusive of passengers): 47 tons
Average load for a day's working (exclusive of passengers): 44 tons
Coal consumption: 29 lb. per mile
Oil consumption: 0.47 pints per mile

Brighton and Worthing Service.

Maximum load (exclusive of passengers: 91 tons
Average load for a day's working (exclusive of passengers: 68½ tons.
Coal consumption: 35 lb. per mile
Oil consumption: 0.59 pints per mile

The 'A' class trains are operated by the driver, either from the engine footplate or from the driver's compartment at the remote end of the coach, according to the direction in which the train is moving.

In the case of the 'D' trains, these are operated only from the driver's compartments in the end coaches, the driver transferring to the leading end of the train at the beginning of each trip. In the driver's compartment are placed all the necessary fittings which give him complete control in operating the train.

The fittings consist of a driver's brake valve and DUplex air-pressure gauge for operating the Westinghouse brake, a whistle blown by compressed air and an air valve for controlling the movements of the steam regulator on the engine. In connection with the latter there is an air-pressure gauge, which is a dUplicate of that used in conjunction with the Westinghouse brake also, an electrically-operated indicator, the hand of which moves in unison with the steam regulator handle on the boiler and so indicates to the driver its exact position at any instant. On the engine the steam regulator handle is furnished with a short forked lever at right angles to itself and attached to the fork by means of a slip pin is a connecting rod, the lower end of which attached to a crankshaft placed under the footplate, which in turn is connected to the piston of the actuating cylinder.

The compressed air for operating the steam regulator is taken from the main storage reservoir of the Westinghouse brake. Air from this source passes through a reducing valve into an auxiliary reservoir, at a pressure of 25 lb. per sq. in. lower

than the main; this auxiliary reservoir is in direct communication with one end of the actuating cylinder and maintains a constant back pressure on one side of the piston, keeping the steam regulator closed. One hand of the air gauge in the driver's compartment at the end of coach is also in direct communication with the end of the cylinder by means of a ¼-in. bore pipe and indicates this exact back pressure obtained.

The usual main storage and train pipes for operating the Westinghouse brake are, of course, carried along under the coach from the engine to the brake valve in the driver's compartment, and at this point a branch is led off from the main storage pipe to the air valve for working the steam regulator. A return pipe from this air valve communicates with the actuating cylinder on the engine and, when the valve is opened, admits air from the main storage to that side of the piston which is opposed to the back pressure. When the air pressure so admitted equalises and then slightly exceeds the back pressure, the piston is moved and the stem regulator is gradually opened; on placing the handle of the air valve in the neutral position, the pressure on either side of the piston is maintained in equilibrium and the regulator remains stationary; by moving the handle to the exhaust position, the air is allowed to escape from the pressure side and the back pressure returns the piston to the end of the cylinder and so closes the regulator.

It is sometimes necessary to close the steam regulator rapidly, as for instance when the engine wheels slip on greasy rails etc. To obtain for this purpose a rapid movement of the piston in the actuating cylinder, a quick-release valve is fitted to the end of the main pressure pipe where it enters the cylinder, and by the air escaping through this valve a more rapid exhaust is obtained than if it had to traverse the whole length of the pipe and out through the exhaust port of the air valve in the driver's compartment. The valves used for this purpose are obsolete Westinghouse brake triple valves, which are easily adapted by a slight modification of the ports.

To avoid unnecessary waste of air pressure when operating the steam regulator, a three-way cut-out cock is provided on the engine, whereby the coaches at the rear end of the train from behind the engine can always be out of communication.

When a train arrives at the end of its trip, the driver removes the handles from the brake and regulator valves and takes them with him for use on the valves in the compartment at the leading end, on the return journey. These valves are so designed that they cannot be readily opened in the absence of their handles, and thus all chance of meddling on the part of curious or irresponsible persons is precluded. The fireman has the care of all the work required on the engine footplate, and in the event of any unusual happening, when it might be necessary to stop the train, there is provided an emergency cock in communication with the pressure side of the actuating cylinder, on opening which the air is released and the back pressure closes the regulator. There is a pressure gauge on the engine also which indicates the back pressure in the actuating cylinder, and on the pipe leading to it is placed a test and release cock, by means of which he can reduce the back pressure at times when the main storage pressure has fallen too low to overcome it and move the piston. At the attachment of the connecting rod to the lever on the regulator handle a slip pin is provided, which, having it handle on the head and a hinged end at the point, allows of it being easily and quickly uncoupled in the event of failure in the air-control gear.

Opposite: Unfortunately the same image as has been seen before in connection with this subject but we also now know taken a few years earlier c1908.

It is our intention to feature the various railcars operated by the LBSCR and its neighbour the SECR in the near future.

SOUTHERN TIMES

John Davenport
A lifelong railway interest
Part 2: British Railways days

John Davenport's railway interest was a lifelong passion. Although without doubt a Southern man, his paper records and photographs show he would also travel widely when the chance presented itself both to other regions of mainland Great Britain and also into Europe. He was also a fastidious note taker not as some might, just of engine numbers, but the trains they worked. To this end it must have been that he had access to the relevant working timetables. Certainly he had cultivated relationships with various career railwaymen at all levels and which stood him in good stead s the years passed.

A located file also includes hand written descriptions of some of his jaunts, possibly with a view to submitting these for publication. We cannot be sure if this actually took place but even so what follows was compiled almost 70 years ago so a little duplication may be excused! The title was a simply, 'A Schools Day out'.

'We are accustomed these days to a wide choice of special trains on the min lines from a number of operators. Fifty years ago when steam was normal, the choice was more restricted. The various railway societies rail tours generally to or on branch lines, but the Ian Allan organisation have organised some main line runs, including his recent specials to Swindon Works.

'On 25 April 1954, under the 'Trains Illustrated' banner, a day out was organised from Waterloo to Bournemouth on a two-hour non-stop schedule as had been run by pre-war service trains. From Bournemouth there was to be a trip over the Somerset & Dorset to Bath, with a return south as far as Templecombe and then Up the South Western main line.

Because of their association with the two-hour timing, a Schools class 4-4-0 was scheduled for the train., No 30935 *Sevenoaks* had been requested but was unavailable, so we had No 30932 *Blundells* for the day. This was fitted with the original Maunsell chimney. No 932 had been transferred to Bournemouth in July 1937 and repainted in malachite green in July 1938, working the Bournemouth expresses with similarly painted coaches and sister engines. She was then well qualified for the run.

The load for the day was eight coaches, 262 tons tare, 290 tons full. *Clan Line* enthusiasts should remember that a Schools was only rated as 5P! Leaving Waterloo at 9.20 am, 71 mph was achieved at Weybridge and Woking was passed in 28 minutes and 22 seconds against a schedule of 27 minutes. There was a severe permanent way restriction to 20 mph after Fleet and a signal check at Basingstoke. As a result Worting Junction was passed 'seven Down' in 58 minutes 9 seconds instead of the scheduled 51 minutes.

It got more enjoyable after that. The speed at Micheldever was Up to 70 mph and 82 mph at Winchester so that Southampton was reached in 86 minutes 25 seconds against 83 minutes; nearly 3½ minutes had been recovered.

From there to Bournemouth Central there was a signal check at Redbridge and a slowing at Totton. 67 mph was reached at Hinton Admiral and we eventually stopped in 121 minutes 49 seconds for the 108 miles. Our adjusted running time was definitely under the 120 minute schedule.

With the addition od LMS 2P No 40601 as pilot to Bath (the 'Six Bells Junction website refers to a 'T9' as having been the preferred choice as the pilot engine, but it was not found possible to fit tablet-catching apparatus to any of the available members of the class), we set off with a speed of 60 mph Down Parkstone bank with its interesting curve at its foot. After the 1 in 75 climb of Broadstone bank we stopped for 30 minutes, apparently as we learnt later, due to problems with the level crossing at Bailey Gate. In running time we reached Evercreech Junction five minutes early.

For those that did not know the Somerset & Dorset, the next bit was dramatic. If the Settle and Carlisle is the 'long drag', then this was the sharp shock. For the next eight miles to Masbury summit there were long stretches of 1 in 50 with only a brief respite at Shepton Mallet.

Opposite top: A day observing trains at Woking witnessed D15 4-4-0 No 30465 in charge of a Waterloo – Lymington service in the summer of 1953. Travel was a bit more leisurely then than it is now!

Opposite bottom: In the Down slow line just approaching Farnborough (main) is S15 No 30500 with freight for Southampton Docks. *(Excepting the footplate views, which are John Davenport, the remainder are all images John Davenport / Transport Treasury)*

SOUTHERN TIMES

At the start of the climb the speed was 27 mph but it fell to 15½. After Shepton Mallet the minimum was 24½. We reached Bath Green park on schedule for running time but still 30 minutes late by the station clock.

The plan was for *Blundells* to run light engine to Templecombe to be serviced and made ready for the run back to Waterloo. The special taken back to Templecombe by No 40601 joined by sister engine No 40698. The horrendous start from Bath including the 1 in 50 Devonshire tunnel was completed with 17 mph at the top. Radstock North was passed almost on time in 19 minutes 43 seconds, from where to Masbury summit the lowest speed was 18 mph. After 62 mph at Shepton Mallet, arrival at Evercreech Junction was in 48 minutes 30 seconds against a schedule of 56 minutes.

The pilot was detached and No 40601 took the train to Templecombe. Something then went horribly wrong – after 50 years my notes are a little undecipherable – but it seems to have taken from 5.00 pm to 6.05 pm to go from Wincanton to Templecombe, a distance of 3½ miles. The scheduled arrival was 5.10 pm and departure for Waterloo 5 28 pm. (Modern railtours have similar ancestors.)

The run to Salisbury was not helped by a pws check on the climb to Buckhorn Weston, where *Blundells* left the tunnel at 15 mph. Semley was passed at 33 mph, and with no more than 68 mph through Dinton followed by signal checks, Salisbury was reached in 37 minutes 24 seconds against a schedule of 33 minutes.

After watering and a crew change, we set off on the 'ACE' timing of 86 minutes. On the climb to Grateley, Porton was passed in 10 minutes 21 seconds at 40 mph. Downhill produced 72 mph before Andover, passed in 23 minutes 9 seconds, but Uphill again to Hurstbourne caused a fall to 49

Opposite top: Still at Woking, H16 No 30519 is arriving at Woking with the 8.19am freight from Feltham, August 1953.

Opposite bottom: In the country and fortunately John recorded the number of the engine which would otherwise be obscured by the duty number. We learn it is No 30793 *Sir Ontzlake* Salisbury bound on the west side of Worting flyover. The unusual route of the signal wire from Worting Junction signal box may be noted. 8 August 1959.

This page: No 30854 *Sir Martin Frobisher* departing Southampton Central for Bournemouth. John had just had a footplate ride from Waterloo on this engine in charge of Driver Charham - *who is pictured on the next page.*

Above: Driver Charham on No 30864 *Sir Martin Frobisher*.

Opposite: On 27 December 1961, a landslip involving some 10,000 tons of earth occurred alongside the Down slow line west of Hook station. The slow line itself was suspended over a 40 feet drop over a distance of 50 yards, In addition the sliding earth smashed the arches of a culvert carrying a stream over the line. Damned water flooded nearby meadows and a trench had to be cut to allow the water to drain into a nearby river. Limited services were able to proceed by day with both Up and Down services using the Up slow line on which a 10mph restriction was provided. (Although not confirmed it is almost certain temporary connections were provided and Pilotman working involved.) A number of trains were diverted via the Mid Hants and Portsmouth direct routes although weight restrictions prohibited the 'Merchant Navy' class from the Netley line. John recorded these two images showing No 34018 Axminster cautiously threading its way past the slip working 'wrong road' with a Down boat train. He also recorded Nos 35012 *United States Line* and No *34063 229 Squadron* at Hook awaiting their turn to proceed west. The line of open wagons to the right may well be being used to bring ash and other hardcore to the work site. A full service was not reinstated for approximately four weeks. (January 1961 was not a good month for the Southern Region as on the 22 of the month an earth slip also occurred between Kent House and Penge East meaning Up trains from Orpington to Victoria had of necessity to be diverted via the Catford loop.) *Ed - Writing this at the end of February 2023, the railway between Hook and Basingstoke has only just been fully re-opened after a similar earth slip.*

mph. That, fortunately, was the end of the dull bit of the day. By Oakley speed was Up to 66 mph. Worting Junction was passed at 70 mph in 39 minutes 5 second; the schedule was 39 minutes. What followed is best shown in table form:

Distance from Salisbury		Schedule	Actual	MPH
35.9	Basingstoke		41.06	76
41.5	Hook		45.33	78
43.9	WInchfield		47.33	73
47.3	Fleet		50.21	69
50.5	Farnborough		53.09	68
55.7	Brookwood		57.40	78
57.1	Woking Junction	62 min.	60.07	80
62.4	West Byfleet		62.30	79

SOUTHERN TIMES

64.6	Weybridge		64.28	76
69.3	Esher		68.14	75
70.4	Hampton Court Junction	70 min.	69.09	72
76.5	Wimbledon		74.29	64
79.8	Clapham Junction	79 min.	79.28	35
83.7	Waterloo	86 min.	85.51	-

There had been a signal check at Earlsfield which accounted for the slight loss of advantage on the schedule at Clapham, nevertheless even-time had been obtained by West Byfleet, and the average speed for the 37 miles from Worting Junction to Hampton Court Junction was 75 mph.

The crew for this brilliant run came from Salisbury. Considering that the Schools were never very regular engines on that line and certainly not on the schedule, it was a remarkable effort. Maybe they thought it they treated *Blundells* like a 'King Arthur' it would work. It certainly did.'

Above: The Ian Allan Somerset and Dorset special referred to in the text. No 30932 *Blundells* after arrival at Bath on 25 April 1954. Might this have been the only time a 'Schools' traversed the S&D?

Opposite top: Driver Cambury with No 35020 *Bibby Line* on the Down 'ACE' John rode on the footplate between Waterloo and Salisbury. The position of the regulator handle may be noted!

Opposite bottom: A summers day sees No 35002 *Union Castle Line* passing the closed Bramshott Halt in June 1958 with the 2.30pm Down Waterloo to Bournemouth.

A final instalment from John Davenport's records 'Observations at Farnborough' will follow in Issue 6.

ISSUE 5

SOUTHERN TIMES

Left: This time it is Driver Cutting on No 35018 *British India Line* 'somewhere' between Southampton and Basingstoke on 27 April 1960.

Right: Headquarters Inspector Plummer. No details but certainly on one of John's footplate trips. At this point it may be appropriate to bring in an email recently received from Keith Dawe (former Eastleigh fireman). Keith has had his own excellent book of footplate reminiscences recently published but contacted ST as a result of John Davenport's first article as appeared in ST4. Keith writes, '.... my attention was instantly drawn to the lower picture on page 47. Inspector Danny Knight clearly making himself comfortable in the poor fireman's seat. Oh dear, how I used to hate it when we had visitors on the footplate! It didn't happen very often, but was a real pain when it did. I have previously recounted to you the week we had Mr Townroe taking firemen for their final practical driving exam, but he was naturally stood behind the trainee driver and thus out of my way. But we also once had one of these official footplate pass visitors which apparently had to be accompanied by an Inspector, who had nothing better to do other than to sit in my fireman's seat while the guest stood behind the driver. It was certainly a case of "two's company; three is a crowd and four is just a pain in the neck" A busy fireman does not have a lot of time to be sitting down, but he does like to lean out of his side widow from time to time, just to get a breath of cool air rushing by and check the injector is picking up correctly. Also on a Bullied, both live and exhaust injector valves are under the fireman's tip up seat - which you need to tip up to gain access. Very difficult when you have a lump of inspector sat on it! Just seeing that picture brought it all back with a smile. (I am sure he could have meant John!)

And from the mainline to the branch. A self propelled p/way trolley at Whitstone and Bridgerule in July 1954.

1923 a century on.

Marking centenaries and anniversaries is not easy from an editorial perspective. In essence exactly what does one commemorate, or in the case of a line closure, commiserate?

Should anniversary recollections also be restricted to routes – and / or stations – or should it also include locomotives new / old, rolling stock, signalling schemes, electrification; the list is almost endless. We are conscious also that a reader with a specific knowledge or interest may well feel disconnected if a specific anniversary important to them is apparently ignored.

All this perhaps subjective perspective pales into insignificance with one anniversary we may mark in 2023. The centenary of the birth of the Southern Railway on 1 January 1923. An amalgam of the three major operators in the south of England, the South Eastern & Chatham, The London Brighton & South Coast, and the London & South Western railway companies.

Government legislation, notably the Railways Act of 1921, had dictated that the 120 or so individual railway companies previously in existence, be formed into four large groups. This had become

The Southern Railway applied for a formal coat of arms in 1938 but it was not granted until March 1946. The arms include allusions to London (sword), Dover (a leopard's head), Southampton (a rose) and Brighton (a dolphin); the blue wave represents coastal areas and maritime operations. On the crest had been intended a Bulleid-Firth Brown locomotive wheel (although as seen as spoken wheel is shown) with the flash alluding to the extensive electrification of the Southern Railway system. The sunburst relates to the Southern Railway slogan 'South for Sunshine'. The two edge supporters are the Red dragon (London) and White Horse of Kent (each resting a foot on another railway wheel). Exactly how much use was made of the crest is not known considering is was not authorised until less than two years before nationalisation. Certainly it never appeared on locomotives or rolling stock where the simple heading 'Southern' was used for the 25 years the company was in existence from 1925 to 1948.

SOUTHERN TIMES

necessary due to shortages of staff and arrears of maintenance in consequence of WW1. It was anticipated that with less competition for the same traffic and an amalgamation of resources, a more efficient means of operating might be found. Using the same rational, the argument for nationalisation post WW2 also becomes clearer.

After the passage of time, it is also difficult to imagine the upheaval Grouping must have been to staff at all levels. The individual previously the small cog in a small wheel, was now potentially reduced in status further.

Some names, and their origins of course became embedded in history. Messrs Walker and Maunsell for example taking the top posts in their respective fields whilst in other senior positions again it was one man who was selected perhaps to the chagrin of his opposite numbers.

We are not told in the previously published histories whether duplication of rolls also now meant redundancy, or was excess capacity resolved through natural wastage. These issues would not have been unique to the Southern although as the smallest of the 'Big Four' there would presumably have been less to deal with.

We might also consider what, apart from individuals, each of the three constituents brought to the table. So far as the SECR was concerned it appears this was locomotive policy; the ideas of Mr Maunsell to reign supreme for the majority of the existence of the SR. That from the LBSCR is more difficult to define in simple terms, but we might state the use of the Westinghouse air system as a standard for pull-push working (as indeed described earlier) and the ambition at least for electrification on what was a main line; London to Brighton, even if when such did eventually appear it was not using overhead wires as had been anticipated prior to 1923.

The LSWR was the largest participant in almost every area. It was on former LSWR territory at Waterloo that the general offices of the enlarged company were now established and whilst it would unfair to say that LSWR influence quickly spread east, there is no denying that as time passed so more former LSWR stock was seen across the whole system than perhaps examples from the other parties. LSWR third-rail electrification was also the one that was followed, a legacy which today we may regards as more of a millstone but that was far into the future into 1923 and the newly established Southern Railway cannot in any way be held responsible for what was considered right at the time.

Personalities also come to the fore and here again it is the chiefs – and prospective chiefs that are the best known. Sir Herbert Walker from the LSW taking on the role General Manager whilst a rising star from the SECR, Eustace Missenden would assume the role later.

The intention of this very brief piece was never intended to be more than an acknowledgement of the passing of a hundred years. We would be very grateful if readers might add their own views and opinions as to the amalgamation

Did you get them all right…?

(Page 80 of Issue 4.)

1 - Wroxall

2 - Leysdown

3 - Hayling Island

4 - Battle

Another teaser on page 77.

Stephen Townroe's Colour Archive; Adams and Drummond Part 2

This is the second selection of material from the Townroe archive to feature the designs of Messrs Adams and Drummond. We should not forget that some of the examples seen had already provided half a century of worthwhile service when photographed and their remaining years of service were also rapidly drawing to a close. None would survive to the end of steam with just a single example subsequently taken into preservation.

Adams 0395 0-6-0 No 30577 (renumbered from SR No 3441) on the main line near Woking. One of 18 engines of the type inherited by British Railways, this example was based at Guildford for most of its BR life and was to be seen literally pottering around on local freight workings until withdrawn in February 1956. The square Adams design cab windows are a giveaway to its origins.

Don't forget, electronic copies of the S C Townroe colour images are available via the Editor.

SOUTHERN TIMES

Opposite top: The former SR 'Royal' engine (which we will see again later), T9 No 30119 seen here on a less exalted duty at Southampton Central; a summer Saturday Southampton to Dorchester working in August 1951. It was thus returning to its home shed of the time. As to why No 119 (30119) came to be regarded as the Royal engine is not certain although we do know that for part of its life it reposed under a tarpaulin at Nine Elms. Its society position was not enough o save it from scrap when diagnosed with cracked frames and withdrawn at the end of December 1952.

Opposite bottom: Stephen Townroe appears to have taken the opportunity to make several visits to the Meon Valley line including as seen here a visit to Droxford in February 1955 where 700 class No 30350 was waiting for the single line section to clear with the arriving pull-push working before continuing its leisurely journey north to West Meon and Alton. The Meon Valley line closed soon after and was severed above Droxford.

Above: A 'Greyhound' (another T9) reduced to sniffing around various goods yards for traffic compared with its express duties of decades earlier. This is No 30287 crossing the junction with the Sprat & Winkle to Andover at Kimbridge Junction with a Salisbury to Eastleigh pick-Up goods.

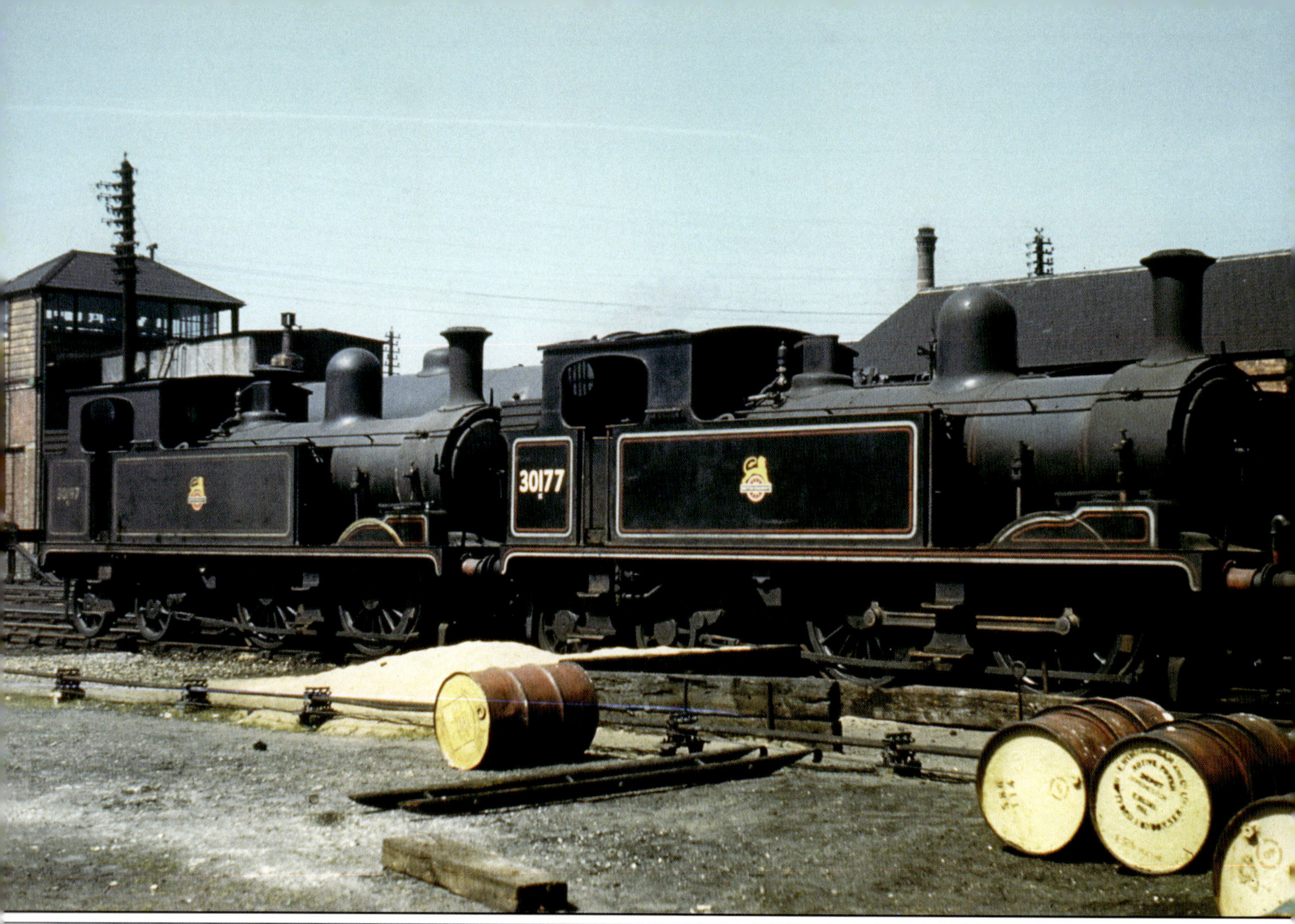
30177

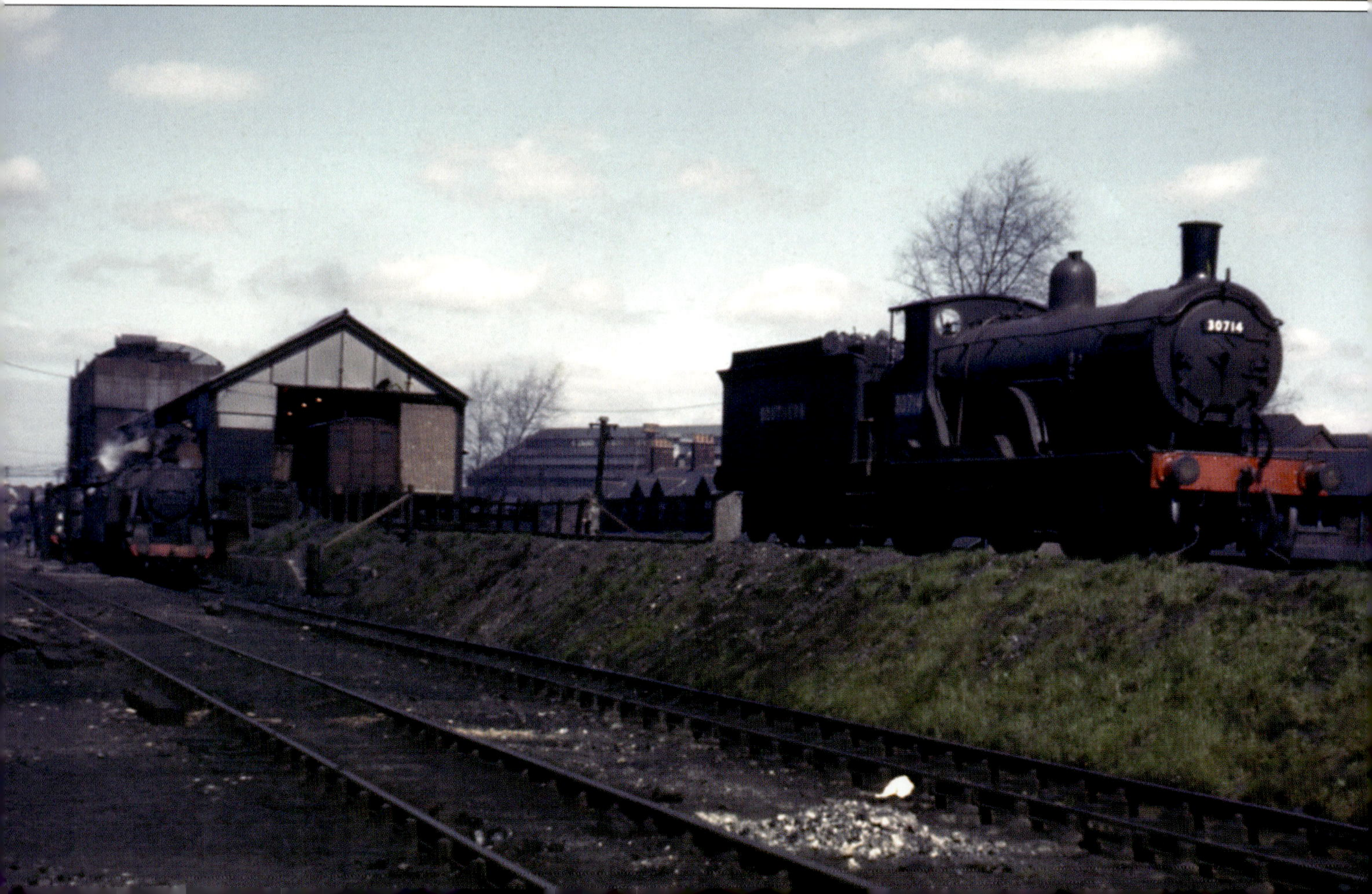
30714

Opposite top: Dorchester shed in May 1952. Two 'mainland' O2s are visible, Nos 30177 and 30197 together with in the background the tall LSWR signal box. Historically a number of Southern trains would run only as far as Dorchester shunting taking place to allow through passengers and stock to continue to Weymouth via a connection to the Great Western route. Had the Victorian ideals come to fruition, Dorchester would have been a through station on a westbound route to Exeter but this never materialised.

Opposite bottom: Exactly what No 30715 (another T9) was doing on the coal stage road at Eastleigh in 1948 is not certain, it was not the normal type of engine one would associate with shunting the gradient with coal wagons. Possibly it was simply out of the way but in its position gives a good example of a Southern tender attached to an engine which is otherwise in BR livery.

Above: No 30119 and this time with the tender at least a bit cleaner than when seen at Southampton Central. The engine seen emerging from Bincombe Tunnel between Dorchester and Weymouth in May 1952. The hardcode is for a light engine on what was a WR line rather than the more usual SR route code.

SOUTHERN TIMES

Above: Rural setting. Tisted on the Meon Valley line with another T9, this time No 30732 seemingly caught in the middle of a shunt move. The running line dipped at this point towards Farringdon and Alton, hence the yard head shunt is at a higher level on the right. Strange that the T9s, built as passenger engines do not seem to have been used for passenger work on the line – they were used on branch passenger services elsewhere. Instead it was a goods class, the 700 design, that did the passenger turns. Housing now occupies most of the former goods yard at the station. February 1955.

Opposite: Another Adams type – again the square front spectacles, No 30582 at Eastleigh in 1949 and fresh from overhaul. It is alongside the office building at Eastleigh and will no doubt soon return to Exmouth Junction and its more usual haunt of the Lyme Regis branch from Axminster.

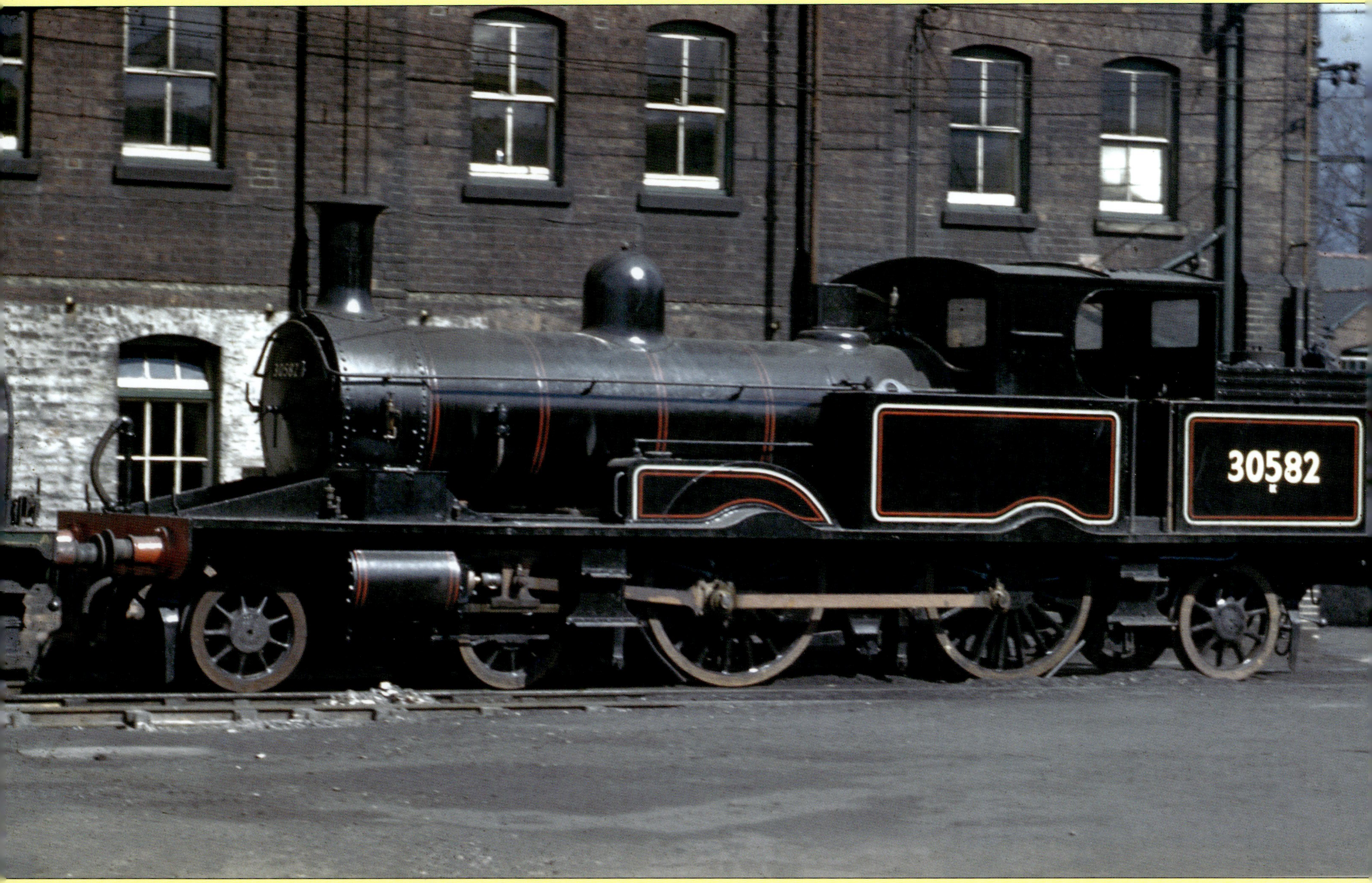
30582

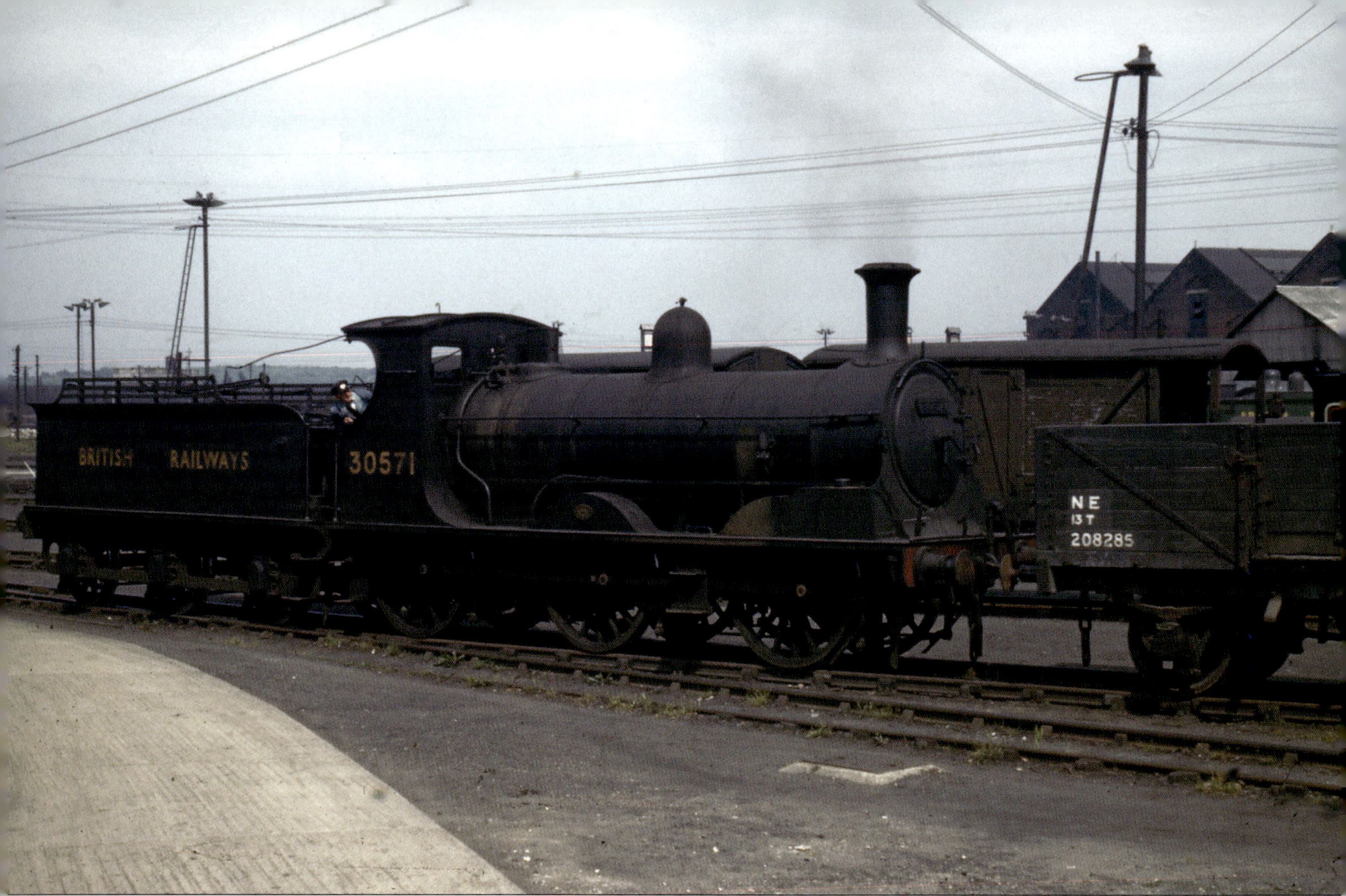
BRITISH RAILWAYS
30571
NE
13 T
208285

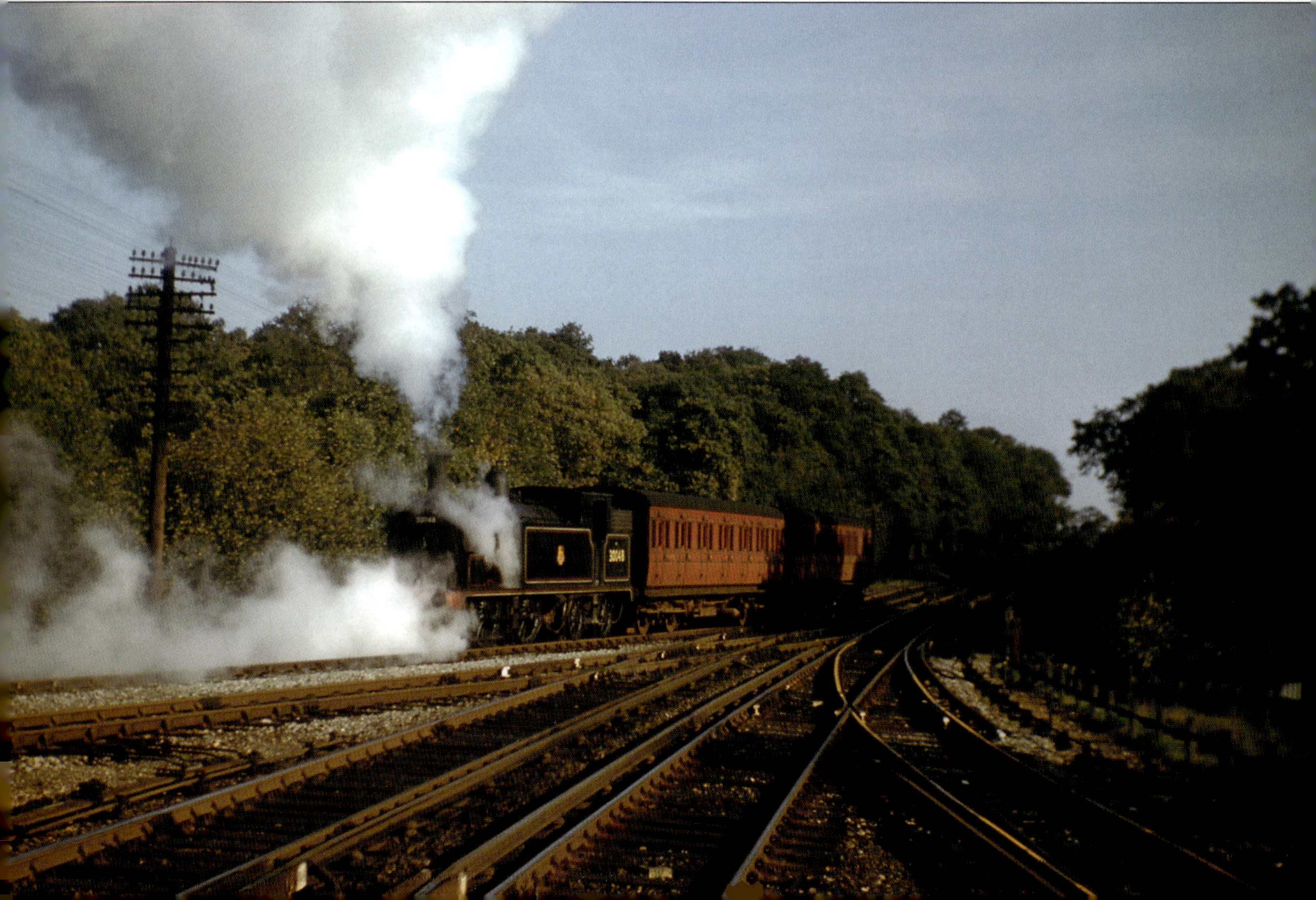

Opposite top: It would be tempting to think No 30571 was seeing out its final days as works shunter at Eastleigh in 1950, but it still had a further three years of life ahead and moved to more general duties at Feltham in 1951. The 0395 type were unusual in having a sloping front to the smokebox.

Opposite bottom: A Guildford to Horsham pull push working sets off from Christs Hospital sometime in the summer of 1955 behind M7 No 30048. From the amount of steam escaping perhaps there had been a layover at the station meaning the need for the cylinder drain cocks to be opened, or else the engine is perhaps not in the best of condition.

Top: Drummond K10 still as SR No 345 eking out its last duties in 1949 with a an Andover train via the now long closed Stockbridge route sometime in 1949. Allocated No 30345, the number was never carried as the engine was withdrawn later in the same year.

Bottom: No date, but we know it is at Eastleigh and probably 1948/9. This particular M7 still displays Southern colours but looks well with the British Railways yellow lettering and number. All the class would eventually go into BR black; some still lined out. A question that arose recently was what engines (classes) had their wheels painted green in SR days and was this applied at all three works? Some comments would be welcome as a study of contemporary images fails to provide a consistent answer.

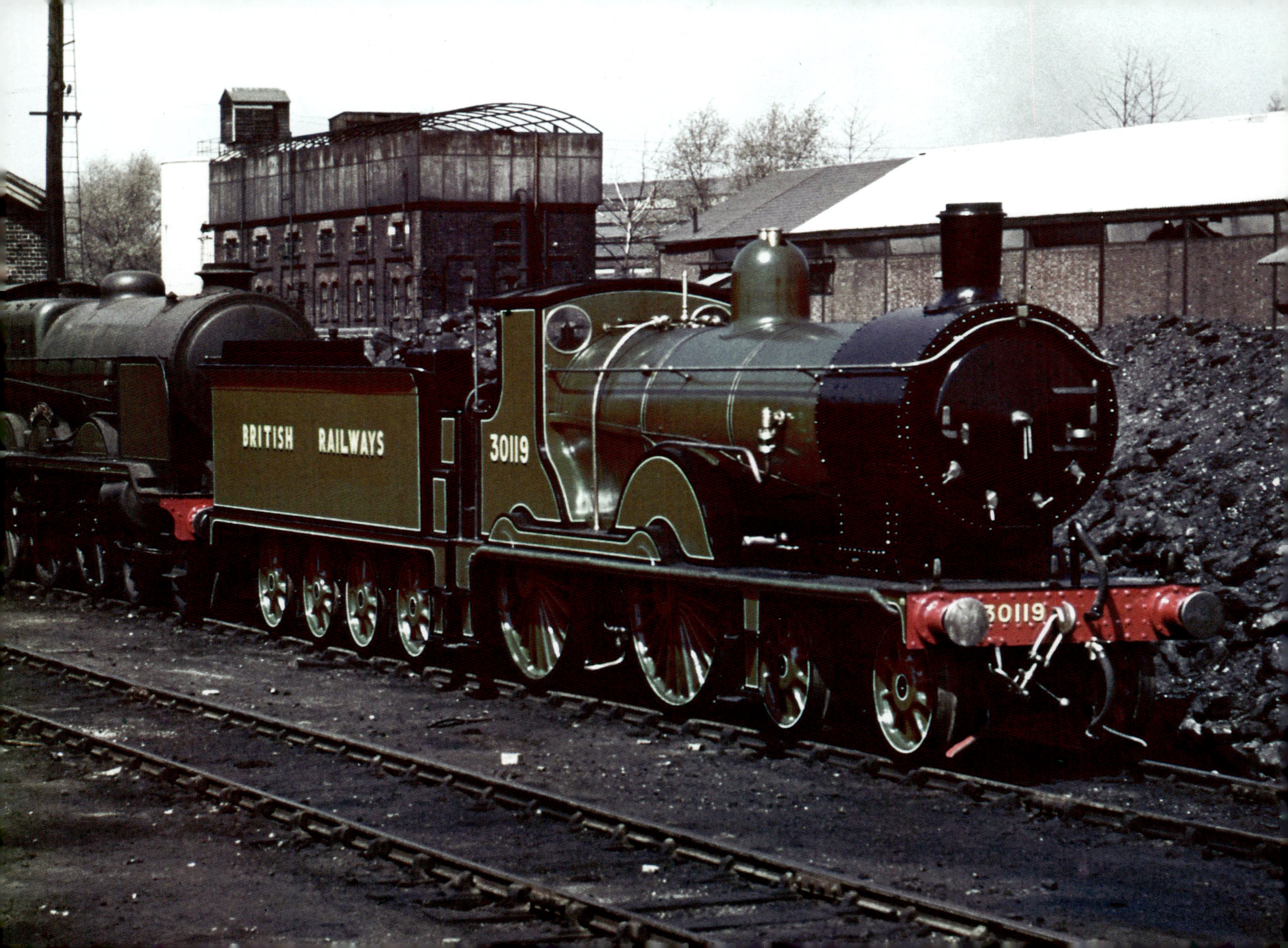
BRITISH RAILWAYS
30119
30119

Opposite: Our final view of No 30119 again, this time pristine at the rear of Eastleigh in 1949 and with an N15X behind. Proof of the importance of this particular machine may be gleaned from the polished smokebox door clamps and coupling. (Yes, the wheels are green on this engine.) Pseudo- Southern livery but with BR lettering and yet the front number is painted on the buffer beam rather than identified with a cast plate. The emergency coal stock alongside is perhaps best described as being of dubious quality, hence stacking has not been possible.

Right: And speaking of coal, the fireman on No 30479 is attempting to level his load of coal whilst precariously balanced on the narrow framing and hanging on with just one hand. Modern day Health and Safety would have a field day and yet such practices and far worse were commonplace on the steam railway. The view was taken at Eastleigh in June 1957 and a bit unusual in not displaying any BR emblem on the side tank.

Bottom: The semi roundhouse that was Guildford shed and turntable, no date. Guildford was almost a steam oasis within the electrified network although steam was an essential component for services to Horsham, Redhill, Aldershot and Reading. Visible are B4 No 30089, acting as shed pilot, 0395 No 30578 and a 700 which, according to SCT's notes, is just coming on to shed after having worked a local freight.

Next time: Visiting the Isle of Wight.

BRITISH RAILWAYS – SOUTHERN REGION

RIPE CROSSING

No. 1304

- 1 – WICKETS
- 2 – GATE LOCK
- 3 – GATE STOPS

GLYNDE ← UP / DOWN → BERWICK

SIGNAL BOX

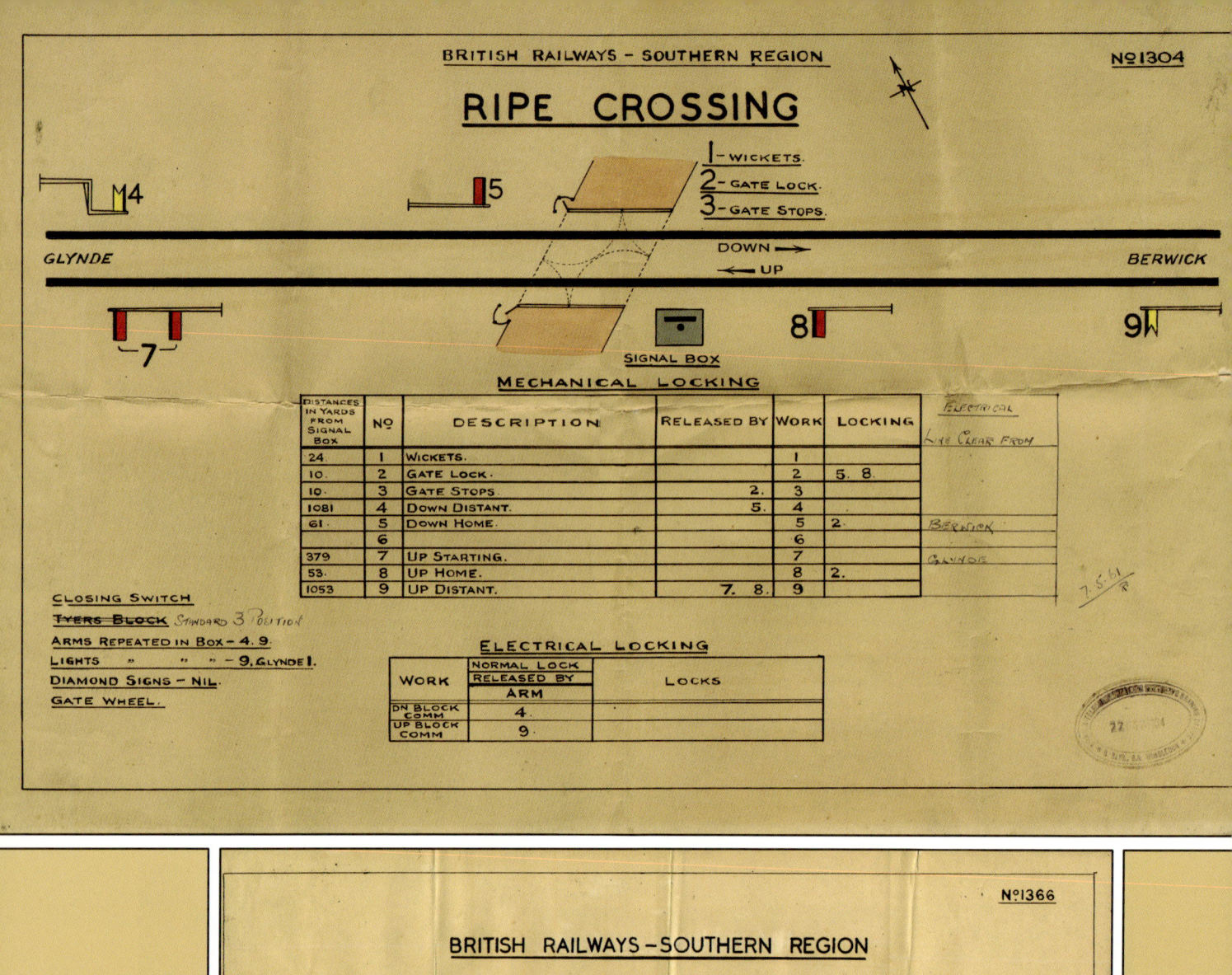

MECHANICAL LOCKING

DISTANCES IN YARDS FROM SIGNAL BOX	NO	DESCRIPTION	RELEASED BY	WORK	LOCKING	ELECTRICAL LINE CLEAR FROM
24	1	Wickets.		1		
10	2	Gate Lock.		2	5. 8.	
10	3	Gate Stops.	2.	3		
1081	4	Down Distant.	5.	4		
61	5	Down Home.		5	2.	BERWICK
	6			6		
379	7	Up Starting.		7		GLYNDE
53	8	Up Home.		8	2.	
1053	9	Up Distant.	7. 8.	9		

CLOSING SWITCH
TYERS BLOCK Standard 3 Position
ARMS REPEATED IN BOX – 4. 9.
LIGHTS " " " – 9. GLYNDE 1.
DIAMOND SIGNS – NIL.
GATE WHEEL.

ELECTRICAL LOCKING

WORK	NORMAL LOCK RELEASED BY ARM	LOCKS
DN BLOCK COMM	4.	
UP BLOCK COMM	9.	

BRITISH RAILWAYS – SOUTHERN REGION

WILMINGTON CROSSING

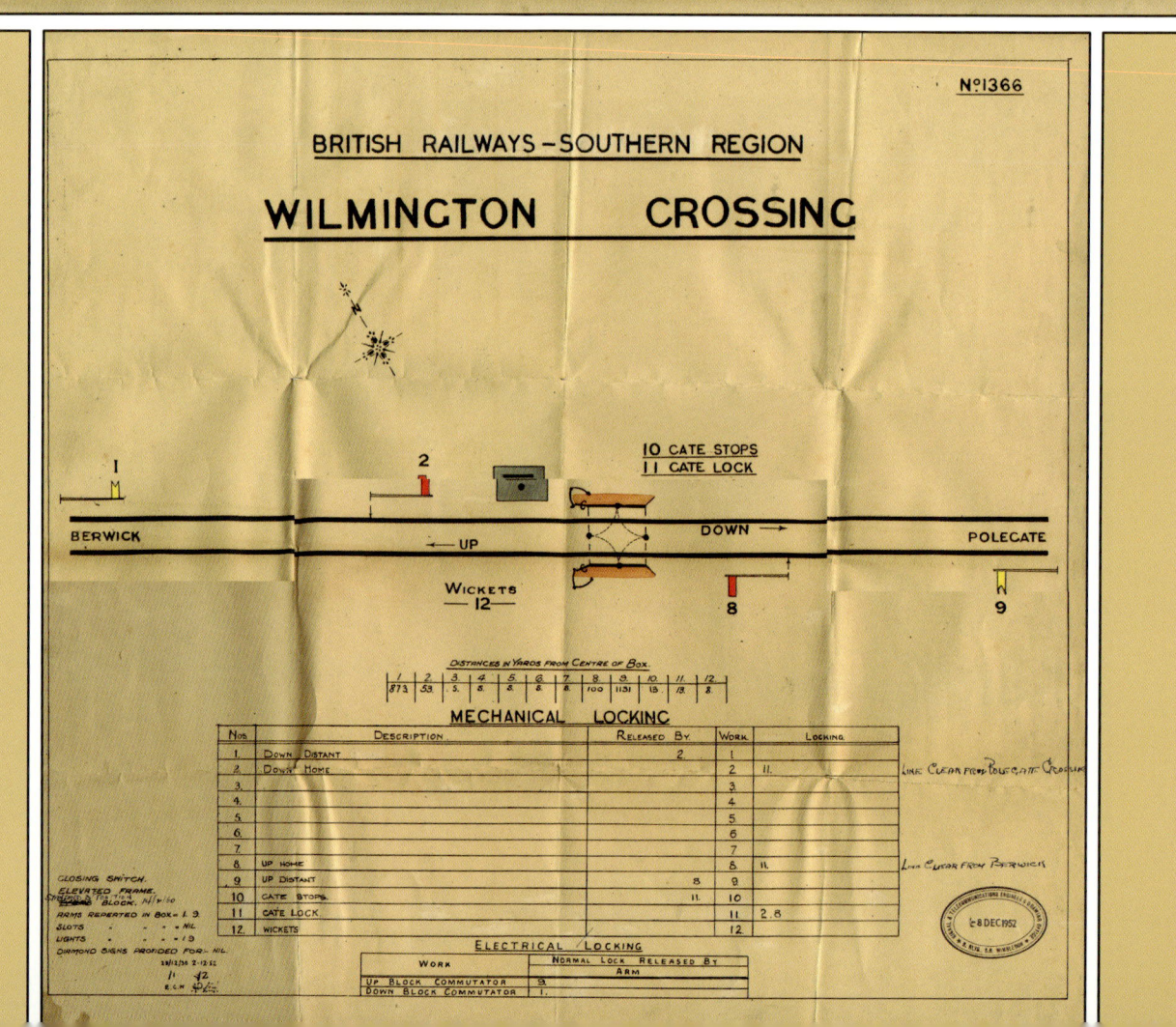

No. 1366

10 GATE STOPS
11 GATE LOCK

BERWICK ← UP / DOWN → POLEGATE

WICKETS 12

DISTANCES IN YARDS FROM CENTRE OF BOX.

1	2	3	4	5	6	7	8	9	10	11	12
873	53	5	5	5	5	5	100	1151	13	13	8

MECHANICAL LOCKING

Nos.	DESCRIPTION	RELEASED BY	WORK	LOCKING	
1	Down Distant	2	1		
2	Down Home		2	11.	LINE CLEAR FROM POLEGATE CROSSING
3			3		
4			4		
5			5		
6			6		
7			7		
8	Up Home		8	11.	LINE CLEAR FROM BERWICK
9	Up Distant	8	9		
10	Gate Stops	11	10		
11	Gate Lock		11	2.8	
12	Wickets		12		

CLOSING SWITCH.
ELEVATED FRAME.
TYERS BLOCK, N/T 10
SLOTS – NIL
LIGHTS – 13
DIAMOND SIGNS PROVIDED FOR – NIL

ELECTRICAL LOCKING

WORK	NORMAL LOCK RELEASED BY ARM
UP BLOCK COMMUTATOR	9.
DOWN BLOCK COMMUTATOR	1.

Two level crossings

A recent discussion with former career railwayman Brian Wheeler (about non-SR topics) revealed a pair of most interesting signal box (or should that really be gate box?) diagrams that have survived even if it is many years since mechanical signalling was present at either location. Both are on the line from Lewes to Polegate and located either side of Berwick station. (Other level crossings also existed in the area.)

The two are interesting as they provide for a simple explanation, and it is to be hoped understanding also, of the principles and associated locking of mechanical signalling.

Starting with Ripe it will be noted that for a train to proceed from Glynde to Berwick, the starting signal No 5 must be clear (what is known as in the 'off' position'). On the diagram it is shown 'on' (at stop) and indeed this was the default position for a mechanical signal; the signal remains 'on' unless it needs to be cleared for a train to pass. (Colour light signals operate slightly differently.)

Now refer to the Mechanical Locking chart beneath and it will be seen that in order to pull No 5 two things must be present. Firstly (last column) 'line clear' must have been received from Berwick, in other words Berwick can accept the train and has turned the block indicator for the Down line to 'Line Clear'. In addition, lever No 2, the Gate Lock, cannot have been pulled. So with lever No 2 'normal' in the frame, and 'Line Clear' having been received, signal No 5 may be pulled allowing a train to proceed. This in turn will free lever No 4, the Down distant signal which is now also free. The locks on the level crossing gate, No 2, and the Gate stops (No 3), similarly locked until No 4 and No 5 are restored to normal with the train past the crossing. Lever No 1 controls the separate wicket gates on either side, which are under the control of the signalman but which may be unlocked at any time.

In the Up direction, the situation is similar although it will be noted there are in fact stop signals on two separate posts (No 7 having co-acting arms.) Accordingly No 8 is classified as a Home signal, and No 7 a Starting signal. So many times we have seen all red arms referred to as starting signals – wrong – it depends entirely where they are placed as to their designation; outer home, inner home, starting, advanced starting, branch home, branch starting, etc, etc. according to local needs. Note also Nos 8 and 7 must both be in the 'off' position before No 9 may be pulled. No 7 has two arms provided purely for visibility and should not be confused with much earlier times when two stop arms on the same post was the same as a latter day junction signal. The early meaning being the top arm for the line to the left and the lower arm for the line to the right.

The rest of the information should be self-explanatory and not require explanation but do take note of the fact the two distant signals, both more than 1,000 yards from the signal box, have the position of their respective arms repeated by means of instruments on the block shelf. Similarly the lamps in signal 9 and also Glynde No 1 (the latter not shown on the diagram) are also repeated.

It was the signalman's responsibility to look at the position of the various signal arms and confirm these had responded to the lever movement. If the signal was obscured, by a structure or by its distance from the signal box, then a repeater was provided.

The gates are operated by a wheel in the box and it will be noted in the position where the signal box is drawn there is a black line with a dot underneath. The former indicated the position of the frame – ie forward facing – with the dot as the centre point of the signal box from which all measurements were taken.

Finally, one vital point is that the signalman was able to return any of his signals to danger at any time should an emergency arise.

The second example is that of Wilmington (please ignore the apparent offset shape of the signal box). With more levers this might at first glance appear to be more complicated but in fact the reverse is the case as there are in fact just seven working levers with the remainder spare.

Both the boxes described were true block section boxes as well, although as each also had a block switch they could be switched out of section if required. Otherwise they needed to be manned when the level crossing might be required for other than pedestrian use.

Each would survive until late 1965 when automatic half barriers were installed operated by track circuits and the presence of an approaching train.

SOUTHERN TIMES

CLOSING OF HOLLAND ROAD HALT

On and from MONDAY, 7th MAY, 1956, the passenger train service stops will be withdrawn and HOLLAND ROAD HALT CLOSED.

Alternative bus facilities in the area are provided by Brighton Hove and District Transport which operates a frequent service between HOVE and BRIGHTON stations.

Further information may be obtained from the Station Masters at Hove (Telephone Hove 32002), Brighton (Telephone Brighton 26211), or from the District Traffic Superintendent, British Railways, Redhill, Surrey (Telephone Redhill 3361)

Enquiries regarding bus services should be addressed to the Brighton Hove and District Omnibus Co., Ltd., Conway Street, Hove 3 (Telephone Hove 31002-3-4), or Brighton Corporation Transport Department, Lewes Road, Brighton 7 (Telephone Brighton 26141-2-3-4-5)

SOUTHERN

Holland Road Halt

Holland Road Halt had a life of just over 50 years, opening in 1905 and closing in 1956. Its raison d'être was to provide additional rail access to the public in the hope they would be tempted back to (or remain) with the railway for local travel as against the electric tramways recently opened in the area.

In that respect the LBSCR were no different to other railways around the country facing similar competition although locally the railway's response was with the use of single coach steam railmotor vehicles rather than electric traction. Other similar local stopping places opened at the same time and for the same purpose were at Dyke Junction, Fishersgate, Bungalow Town and Ham Bridge.

The halt was located to the west of the site of the original Hove station (1840-1880) but east of the later Hove station. Access was by means of steps leading down from the west side of the overbridge at the west end of Holland Road. (There was also an unconnected goods yard with the same name which lasted until 1971.)

The cheapness of construction was exemplified with timber platforms, although at the time of closure it was noted part of the eastern end of the westbound platform at least had been replaced at some stage with a more robust construction using standard Southern concrete supports and a slab subface.

Opposite top: The incentive to travel (but ignore – for now – the disincentive of the closure notice which was displayed on the right of the billboard).

Opposite bottom: Towards Brighton. Rudimentary shelters existed on both sides whilst varying styles of lamp and some outside wooden benches were limits of passenger luxuries. Access and exit was only by means of the steps from the road overbridge; some years before disabled access for passengers was considered a priority.

This page: Train services were not infrequent, with a half-hourly electric service plus steam services on and off the Steyning line. Closure came from 7 May 1956 after a life of just over 50 years. The views seen were taken on 23 April and 5 May 1956 and are all from the camera of Gerald Daniels.

No 142 in beautifully lined out livery as befits the late 19th century. The location is not confirmed and whilst most LSWR locomotive images of the period tend to have been recorded at Nine Elms, this one appears to have a whiff of Northam about it.

William Adams '135' class. Nos 135 to 146

History recalls William Adams would be responsible for 118 4-4-0 tender engines to eight differing designs built over a 17-year period. Externally all would have a familiar family outline with outside cylinders but more careful study reveals considerable variation in the bogies, driving wheel diameter, running plates, splashers, fittings, etc.

380 class. 12 engines built 1879 by Beyer, Peacock & Co. In service 1879 – 1925.

135 class. 12 engines built 1880/81 by Beyer, Peacock & Co. In service 1880 – 1924.

445 class. 12 engines built 1883 by Robert Stephenson & Co. In service 1883 – 1925.

560 class. 22 engines, 10 from Neilsen & Co. in 1884, 12 from Robert Stephenson & Co. 1884 to 1887. In service 1884 – 1889.

X2 class. 20 engines built at Nine Elms 1890-92. In service 1890 – 1942.

T3 class. 20 engines built at Nine Elms 1892-93. In service 1892 – 1945.

T6 class. 10 engines built at Nine Elms 1895-96. In service 1895 – 1943.

X6 class. 10 engines built at Nine Elms 1895-96. In service 1895 – 1946.

On this occasion we will look at the second in the list of 4-4-0s, the '135' class consisting of ten engines built to an Adams design for the LSWR by outside contractor.

The initial order was for just six engines placed on 7 January 1880 for £2,685 each. Some time after this the Traffic Department requirement for the following year's summer service was disclosed and as a result permission was obtained to double the order to 12 engines. All were delivered over a three month timeframe from November 1880 to January 1881.

The design was also a further development of Adams' improvements on his predecessor, having for example a steel boiler and provision for the free passage of steam. Externally the proportions were also pleasing to the eye.

The original intention had been for all of the class to be at Nine Elms working trains to Salisbury or

No 136, undated but with a puzzle as well. Notice the cabside number is struck through; might this be an early means of indicating that the engine is duplicated? Notice too the conical smokebox door, round cab windows and tender with coal rails.

L. & S. W. R.

4 Wheels Coupled Bogie Passenger Engine and Tender. Nos (0135-0146)

Engine weighed with 2½" Scale ¾ to a Foot.
Water in Class, and Light Fire. 12 Thus.

140 Lbs Per Sq.In.

Tender weighed with Tank Full
and 5 Cwt of Coal.

Capacity of Tank 2,500 Gallons.

234 Tubes 1¾ Diar Outside.
Heating Surface of Tubes.....1112 Sq.Ft
    Do   Do   Firebox.....111 "
         Total.....1223 "

Grate Area: 17·77 Sq.Ft

	T.C.Q.
Weight Empty.....	16.13.0
Do In Working Order	17.13.0

Weight of Engine per foot run 1·53 Tons.

	T.C.Q.
Total Weight of Engine Empty.....	43.10.2
Do Do Do In Working Order	46.8.0

	T.C.Q.		T.C.Q.
	13.12.2		13.5.0
	14.9.0		14.6.0

40·4 Total Wheelbase
48·10¾ Total Length over Buffers

	T.C.Q.		T.C.Q.		T.C.Q.
	3.19½		5.2.0		5.4.0
	3.19½		8.11.0		9.7.0
	3.19½		8.14.0		

Weight of Tender per foot run..... 1·45 Tons.

	T.C.Q.
Total Weight of Tender Empty.....	15.8.0
Do Do Do In Working Order	26.12.0

	T.C.Q.
Total Weight of Engine and Tender Empty.....	58.18.2
Do Do Do Do In Working Order	73.0.0

Diar of Cylinders.....18"
Stroke.....Do.....24"
Tractive Force on Rails.....7,874 Lbs.

Type 4-4-0.

Southampton. Soon after entering service though, two appear on the Salisbury allocation and a further pair at Northam (Southampton). Bradley does not mention any specific issues relative to their introduction to traffic and it would appear they then entered service without undue difficulty. Maintenance was also favourable, allowing the type to achieve a good annual mileage allied to coal consumption figures far better than previous Beattie engines. If they had a fault, certainly not according to Adams, it was their weight which was frowned upon by the Engineer – a subsequent stipulated maximum weight for new construction imposed was again ignored by Adams.

In general terms, progress as regards railway engineering was rapid at this time and just three years later another 4-4-0 type had emerged, the '445' class. This led to three being moved to Exeter whilst post-1888 and the more direct route to Bournemouth becoming available, three were also sent to the coastal town.

In service the class are reported to have experienced few incidents, although No 144 had a 'coming together' with Jubilee No 547 on 24 December 1890 whilst the former was on Nine Elms turntable. This resulted in No 144 finding itself in John Street, Battersea over the festive season. No 138 also buried itself up to its axles in earth when the embankment gave way near Weybridge two years later.

Modifications were few, all the class fitted with vacuum ejectors by 1888 with the biggest single change affecting No 136 which under Drummond received a slightly smaller diameter boiler, a spark arrestor and associated conical smokebox door, and a cab having round windows at the front. This took place in 1896. Excepting the removal of the spark arrestor and substituting for a conventional smokebox door, it is likely the boiler and cab changes were retained until withdrawal. Drummond chimneys had also been fitted to the engines by 1906.

It was to be just ten years after first entering service and in consequence of the availability now of the X2 and T3 types that major changes were made to duties and the class were noted to be cascaded on to stopping and even branch line duties. For an engine with driving wheels of 6' 7" such duties were not always ideal, the engines divided between Bournemouth – 5, Dorchester – 3, and Nine Elms-3. One of the latter allocation worked to Reading on a regular basis. A decade later and they had been replaced at the depots mentioned by new Drummond K10 4-4-0s which also took over their previous numbers, meaning the 135 class were now on the duplicate list with '0' added to the front of their earlier three digit identification. The type were also now concentrated on Eastleigh working local services, one longer distance turn being to Woking. Official returns show that when aged 32 at the end of 1912, the average mileage of each member of the class was already in excess of one million miles, indicating they were well used.

The introduction though of the various newer Drummond classes meant they were considered superfluous to requirements and the complete class was laid aside in December 1913 pending scrap, this despite several having recently received repairs. ('Stroke of a pen' condemnation is not only something applicable to BR 50 years later.) More sensible inspection revealed five of the class, Nos 0135, 0139, 0140, 0143 and 0144, were in far better condition than five members of the single frame goods currently still working. It was then but a simple matter to effect a swap; the same number of engines withdrawn and the same number remaining in service.

Perhaps a bit unfairly, it has been said the railway is sometimes run by accountants and what follows is certainly a case in point even if it was more than a century ago, for the LSWR Accounts department could not accept this as a simple swap and now insisted the reinstated 4-4-0s take the numbers of the withdrawn goods engines. Accordingly, between March and April the below change took place:

0135 now 0370

0139 now 0307

0140 now 0310

0144 now 0347

0145 now 0312

Three of the engines, Nos 0307, 0310 and 0347, were reinstated at Eastleigh with the other two at Dorchester. Sometime later in WW1 the Eastleigh engines were moved to Bournemouth for local duties.

By March 1917 No 0347 (No 139) was stored at Bournemouth apparently with damaged motion. It is not believed to have worked again but was not officially withdrawn, on paper at least, for a further four years. Also in November 1921, Nos 0312 and 0370 were towed to Eastleigh for scrap.

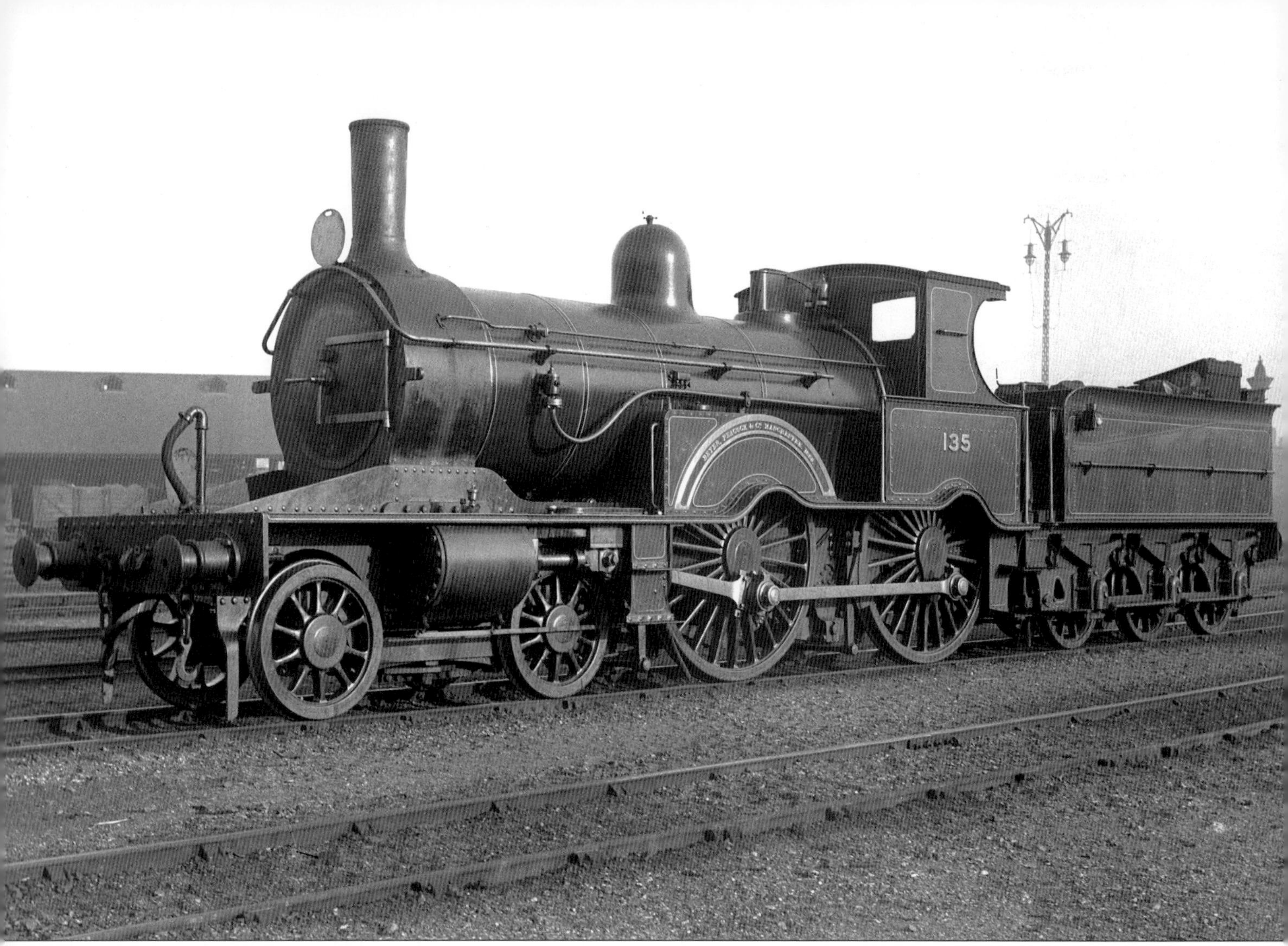
135

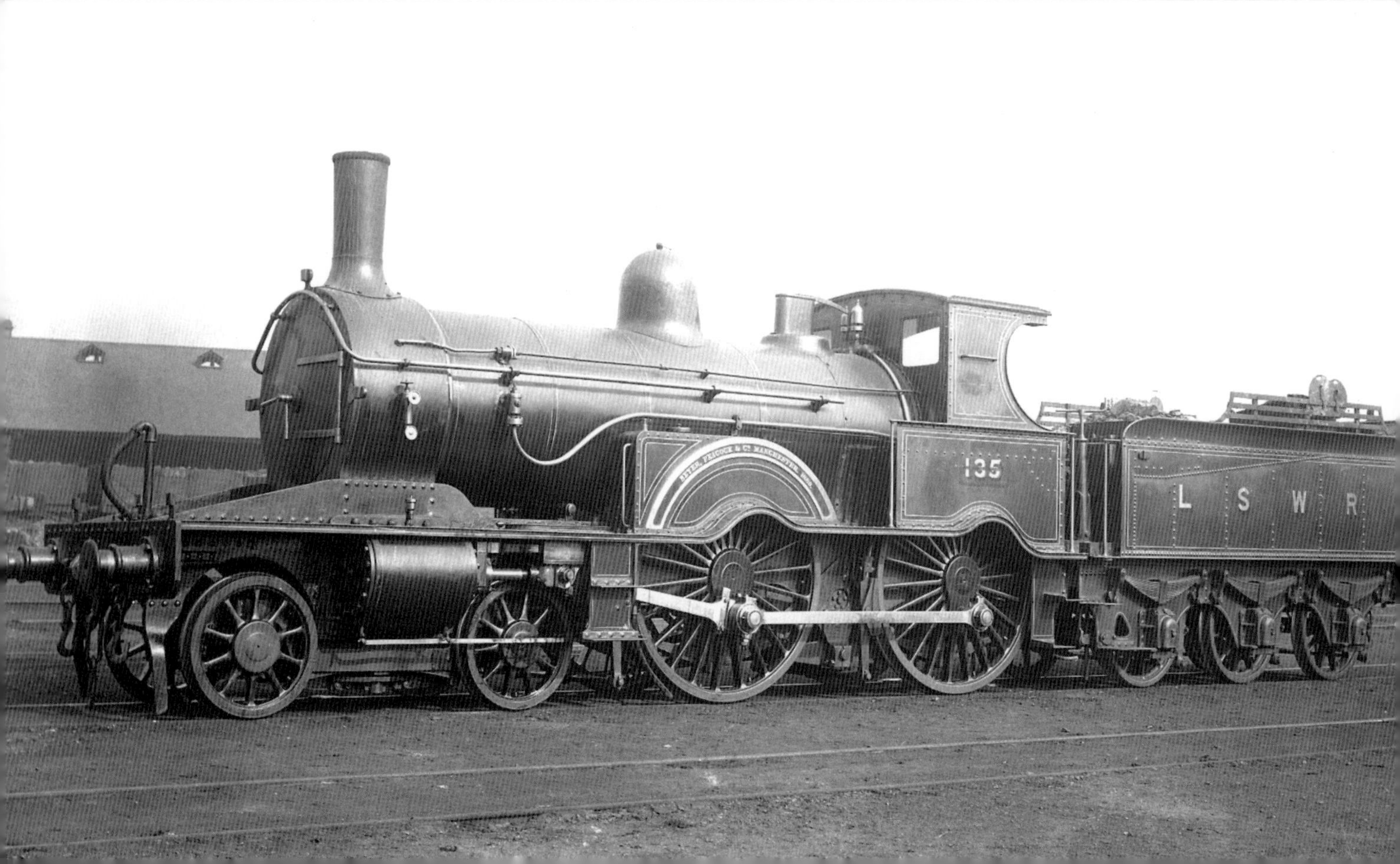
135
L S W R

Opposite top: No 135 displays the large maker's plate around the splasher. Notice too the handrail midway on the tender; its purpose is perhaps doubtful other perhaps than a concession to the cleaners.

Opposite bottom: The same engine from a slightly different pose and with a different tender. The coal rails afford a useful place to store head code discs.

Top: Unusual for the period was to find an image of one of the engines other than on shed. This is No 139 at Basingstoke before the station was rebuilt. To the right is the Great Western terminus. Safety chains may be noted whilst the crew, and possibly the guard as well, were keen to get into the photograph.

Bottom: No 136 again, but this time attached to a tender having different ownership initials.

Under sheer legs but no location given. Open air maintenance was the norm with likely a hand operated - albeit very low geared - hoist.

Again on paper, the last two engines, Nos 0307 and 0310 (the original Nos 139 and 140), entered Southern Railway ownership. Physically both were also at Eastleigh, No 0310 stored at the back of the shed and not thought to have been steamed at all in SR days. It was officially withdrawn in December 1924. No 0307 however was still at work and again according to Bradley on carriage pilot duties, vacuum brake testing of new rolling stock and even local services. It too was eventually taken out of service in December 1924. Because of its continual use, No 0307 (139) may also have achieved the highest mileage for any number of the class, at 1,319,694.

Reference Bradley, Adams locos.

Next time; a break from the LSWR and a look at the LBSCR 'K' class 2-6-0

Southern Region People

We were recently given a small bundle of British Rail Southern Region fact sheets detailing some management appointments in the 1970s. These were prepared as press releases from Waterloo and may well be of interest to readers who perhaps worked with these men, as well as others wanting to know who was in place 'behind the scenes'. In alphabetical order only:

Gordon Graham:

Chief Operations Manager.

3 June 1977.

'New man in charge of running the Southern Region's 5,000 trains a day is Mr Gordon Graham, former Divisional manager of the King's Cross Division of British Rail.

'Mr Graham succeeds Mr Malcolm Southgate's appointment as Deputy General Manager of the Southern Region.

'Mr Graham joined the London and North Eastern Railway in 1939. From 1942 to 1946 he served with the Royal Air Force as a navigator with the Pathfinder Force of Bomber Command. He completed three tours of operational duty during which he was awarded the DFC and subsequently the Bar.

'After demobilisation Mr Graham returned to the railway, working in various posts, principally in the North Eastern Region.

'In 1961, Mr Graham left his position as station master at Ipswich, to be head of Central Timing and Diagramming for the Great Eastern line at Liverpool Street. He was subsequently District Manager at Wakefield, Operations Officer North Eastern Region York, Divisional Movements Manager Sheffield, and in 1970 Movements Manager at Paddington.

'Mr Graham is a Fellow of the Chartered Institute of Transport.'

SOUTHERN TIMES

Left:

Robert Newlyn: Manager,

South Eastern Division.

12 May 1978.

'Mr Robert (Bob) Newlyn, Public Relations and Publicity Officer* for the Southern Region since October 1976, has been appointed Manager of the Region's South Eastern Division. He takes over on 15 May, and succeeds Mr Robert Prescott, who died in February.

'Mr Newlyn has previously served in the South Eastern Division as part of his 37 years on the railways, his appointments ranging from Station Master to senior Divisional Officer. He is a member of the Chartered Institute of Transport.'

(A similar sheet was produced when Mr Newlyn had been appointed as PRO for the SR in 1976. At the time it was stated he had spent all his working life on the Region, latterly in the role of Assistant Divisional Manager of the SE Division. Mr Newlyn took over from Mr R. V. Townsend. Mr Newlyn was the son of a Southern railwayman. He started work as a messenger at Plumstead, then a booking clerk at Falconwood, Station Master at Dover Marine and Dartford, Area Manager at Bromley South and later Ashford, and then Divisional Passenger Officer. In addition he had also worked at Waterloo on marketing plans for passenger services, had experience of freight movement and was concerned in the electrification of the railway on the Isle of Wight.)

Right: 'Second in command on the South Eastern Division,

Mr Robert Powell

came to the Southern Region from the London Midland Region in 1970. Before moving to Beckenham he was Passenger Services Assistant at Waterloo headquarters.'

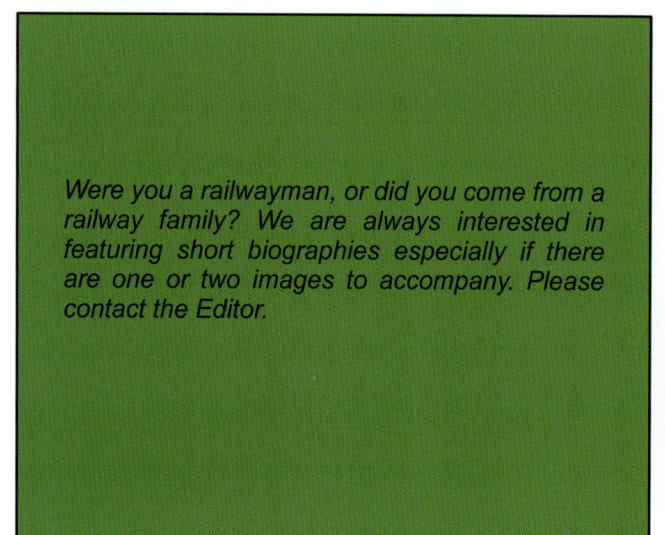

Were you a railwayman, or did you come from a railway family? We are always interested in featuring short biographies especially if there are one or two images to accompany. Please contact the Editor.

'Right:

Movement Manager for the South Eastern Division since 1971,

Mr Brian Hamment-Arnold

has been appointed Regional Management Services Manager at the Southern Region's Waterloo headquarters. In his new post he will be the General Manager's consultant and chief advisor on all matters of productivity – in the fields of investment, manpower, organisation, service quality, marketing policy and management.

Left:

Mr John Palette:

General Manager.

22 December 1976.

'Southern Region's new General Manager under the British Rail reorganisation announced this week is Mr John Palette, General Manager of the Scottish Region. He takes over on Monday 10 January from Mr Robert Reid who is joining the British Railways Board as the member for marketing.

'Mr Palette has spent most of his career on the Western Region apart from nearly two years in Manchester as Divisional Manager and the following six months in Scotland.

'Born in Didcot, he started work as a 14 year old junior clerk in the last war. After serving in the forces, he was selected for a three-year course in railway management. Then followed various appointments in South Wales and London before becoming the Western Region Passenger Manager in 1967.

'Two years later he moved to Bristol as Divisional Manager and then went back to the Paddington headquarters as Assistant General Manager for Movements and Marketing.'

Guildford (and its environs)
Part 2, Colin Martin

Continued from Issue 4.

Opposite top: For our second selection of Colin's images we start with M7 No 30049 at Guildford. The engine is in its final BR livery complete with electrification warning flashes and from the extra pipework is also pull-push fitted. This engine spent from 1951 through to its demise in 1962 based around Horsham, Brighton and Guildford and consequently would have been a regular performer on trains to Horsham and from Brighton on to the Steyning line. No 30049 may just have arrived at Guildford as the headcode is for a light engine via Woking.

Opposite bottom: Lurking around the coaling stage is Q No 30541. Built very much without frills, although the type was fitted with a steam reverser, the 20 engines of the class performed quietly and without fuss throughout their lives. No 30541 was based at Guildford for the final 19 months of its life, ceasing to work at the end of November 1964.

This page, top: The two discs should imply a particular route but what route is not clear; although the tail lamp suggests the engine is in fact propelling an earlier arrival back into the carriage sidings.

This page, middle: From the outlook of the crew, this could also be a propelling move, U No 31623 just north of the station. Notice the engine is right hand drive and also fitted for AWS. The coal perhaps not of the best quality.

This page, bottom: Finally for now (more from Colin elsewhere on the Southern in a later issue), we have Ivatt tank No 41260, then based at Brighton, likely running around its train before returning south again. Originally based at Blackpool and then Fleetwood, it made its way south in 1961 and would remain on the Southern until scrapped in late 1964.

Colin Tuttle - fourth generation railwayman

My family's connection with the railway industry probably began at its outset. My mother's grandfather, Benjamin Broadbridge, was employed by the London, Brighton & South Coast Railway, in the locomotive department at Brighton in the 1850s. It was Ben Broadbridge's second marriage to Mary Ellis that then started the tradition that has carried on down the years.

Four daughters and a son were born from marriage and the family settled at Three Bridges, possibly in a railway company house. The youngest daughter (my maternal grandmother), Annie Marion Broadbridge, married Thomas Merrett, a locomotive fireman, who had lodged with them. The newlyweds initially lived with the Broadbridge family at Three Bridges, but later moved to New Cross, possibly due to Thomas being promoted within the footplate grades. Sadly the career of Thomas was tragically curtailed by his coming into contact with the 6,700v ac overhead traction wire of the 'Elevated Electric', he never fully recovered from the serious injury that resulted. (Family folklore has it he was saved from an even worse fate because of the presence of cork inserts in his boots!)

Even so Thomas and Annie Merrett raised a family, two sons and two daughters. Their eldest son, Thomas (junior), was born in 1896 and achieved 51 years in the footplate grades on the LB&SCR, Southern Railway, and BR(S) respectively. During his career, he was based at New Cross Loco, then Bricklayers Arms, and was finally driving Hastings line DEMUs before his retirement in 1961. Following his father's accident, Thomas (junior) had to become the bread-winner for the family. His younger brother Benjamin, born in 1904, also joined the LB&SCR, but as a booking clerk. His career was in administration, and he rose through the grades to relief station master. Prior to his retirement in 1963, he was Assistant Station Master at Victoria, where his many duties often involved attending to visiting heads of state and royalty.

My mother Isabel, the younger daughter of Thomas Merrett (senior), was born at New Cross in 1906. Meanwhile my father, Albert Tuttle (born at Haywards Heath in 1904), joined the LB&SCR in 1919 as an office boy in the Signal & Telegraph Inspector's office at East Croydon, later transferring to New Cross, where he met and later married Isabel Merrett. Two sons were born to them in the 1930s, and then I followed on in 1946. Albert's career was in railway accounts, and for many years he worked in the S&T accounts section, based at Dundonald Road, Wimbledon. His office later moved to Southern House, Croydon in 1966, where he retired in 1969 after 51 years' service.

My own railway service followed a slightly different path. I spent 43 years with the London Underground (and its later manifestations) in railway signalling, new works and incident management. Not quite the Southern perhaps but still a fourth generation railwayman.

Left: Thomas Merrett (snr) is seen on the right. The identity of the other footplateman is unfortunately not known.

Above: Albert Tuttle.

Nine Elms Coaling Plant
From the 'Southern Railway Magazine' February 1924

The largest locomotive coaling plant in the country has commenced operation at Nine Elms. The principal employed is to hoist the coal into an elevated bunker, from which it is fed by gravity into the engine tenders.

The bunker is a large reinforced concrete construction covering an area about 40 feet by 35 feet and rising to 52 feet above rail level. It is divided so that two classes of coal can be kept separate, and the internal walls are so arranged that each group of chutes (of which there are four groups of two chutes each) is fed, one by one class and one by the other; this makes it possible to load both classes of coal on to the same tender without moving the engine. The capacity of the bunker is 400 tons.

Filling the bunker is accomplished by a side discharge wagon hoist tippler; by this arrangement the wagon on a cradle is lifted up one side of the bunker, and when it reaches the top of the bunker the whole tips sideways. Before actually going into the bunker, the coal passes through a grid, thus preventing too many large pieces of coal being fed to the engine. When the truck has tipped to such an angle that the whole of the coal is out, switches operate which cut the power to the motor, so that there is no risk of overwinding, the controller is then reversed, and the truck returned to rail level where it is automatically stopped. Hoisting, tipping and loading occupies only four minutes.

Any size or type of wagon can be used on this plant, the only fixing requirement being putting on the brake. It is designed for 20 ton wagons. Hoisting the wagon and cradle is done by four wire ropes working over wheels with the necessary gearing and motors on a steel structure on the bunker roughly 90 feet above rail level.

Coal is fed to the tenders by an arrangement of valves and chutes in 10 cwt lots. Opening the first lever fills the measuring chute; the first lever is then closed and the second opened which deposits the coal on to the tender. Eight tons of coal can be placed on a tender in a little under two minutes.

The Nine Elms monolith, operating on the same principle as those on other railways. The Southern Railway eventually built similar (but note, not all of identical capacity) coaling plants at Exmouth Junction, Feltham, Ramsgate, and Stewarts Lane.
R. E. Vincent / Transport Treasury

Electrification to Portsmouth, July 1937
The minutiae behind the scene.

Electrification of the Portsmouth Direct saw services commence on 4 July 1937, the occasion commemorated with due diligence on the ground and within the Southern Railway Magazine for July - where an image of a 4Cor set on the route would certainly have been taken a few weeks earlier. (A further report an illustration was included in the November issue.)

The July issue had the electric train as its front cover and with the following two pages devoted to detailing the new infrastructure and of course the new trains themselves.

Readers of ST will not need reminding of the appearance of a 4 Cor set; the 'one-eyed' look created by the need to display a route indicator with the then standard numerical route / train description instead of a having a window either side at the front. In this respect the 4 Cor sets were unique, for whilst other main line express electric sets had preceded them, Pan / City / Pul / Belle, none had been designed to afford a gangway connection between sets and the Southern was thus faced with an issue previously not considered.

It may be of interest to readers then to reproduce verbatim the wording of a small file of papers recently unearthed which show just how much time and discussion was expended on what nowadays might have seemed an obvious solution. As before with such contemporary documents we have deliberately reproduced these as written and not changed the wording to fit modern day language. A number of questions are also raised which answers are probably now lost in the midst of time.

The first of the papers refers to a meeting of 26 March 1936 held at Eastleigh at which the following were present;
Present

Mr E A W Turbett (Chair)	CME Department
Mr L Lynes	CME Department
Mr G Key	CME Department
Mr F Munns	CME Department
Mr W C Moore	Electrical Engineer for New Works Department
Mr E G Humfress	Electrical Engineer for New Works Department
Mr A E Roberts	Electrical Engineer's Department
Mr C E Finch	Traffic Department
Inspector Shaw	Traffic Department

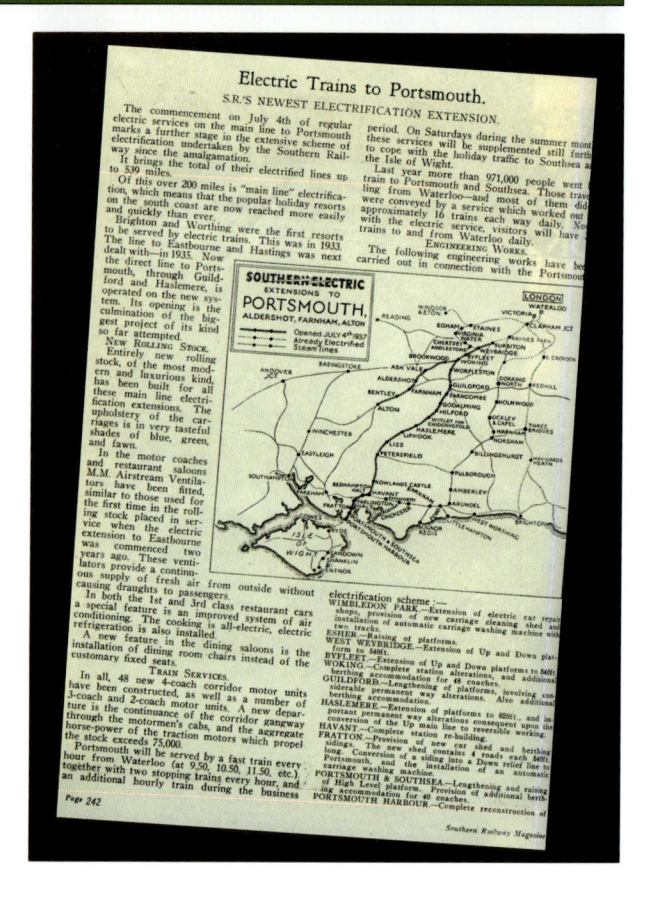

The model motor nose and driver's cab arrangement for the 4-coach units of the Portsmouth Fast Service stock were inspected at Eastleigh and matters arising discussed and the following decisions made:-

Route Indicator. The model was fitted with a standard route indicator with its face parallel with the gangway face and in position shown in sketch 'A' attached (*shown on p 66*).

As the site selected for the route indicator will now be on the flared portion of the nose end, it will be necessary for the coach to be built with projections marked in red on the sketch, to allow for the use of the standard fitting.

The position of the indicator on the model was considered to be too high from the floor level of the coach. It was decided to lower the indicator so that the top of the box is level with the top of the driver's window on the opposite side of the coach, in which position it would be satisfactory.

The 'standard' view of Cor sets. The position of the route indicator dictated a one-eyed appearance giving rise to to nomenclature 'Nelsons'. Popular for many ears they were known to sway at speed especially south of Guildford where a secession of reverse curves that typified the route and so almost accentuated the naval connection. The view was probably taken in advance of the start of the electric service as the front unit is is 4 Res but is temporarily formed with two corridor trailer cars in the centre; presumably the catering vehicles had not been completed at this stage. Mike King comments that the seemingly pedantic attitude to this one item on the SR sets was not unique. The BR Carriage Standard Committee formed in early January 1948 to oversee the production of BR Mk 1 stock looked similarly at absolutely every detail of their construction - ranging from the important like leading dimensions and construction methods/materials, to the trivial including the shape of the soap filler bowls in the lavatories.

SKETCH "A"

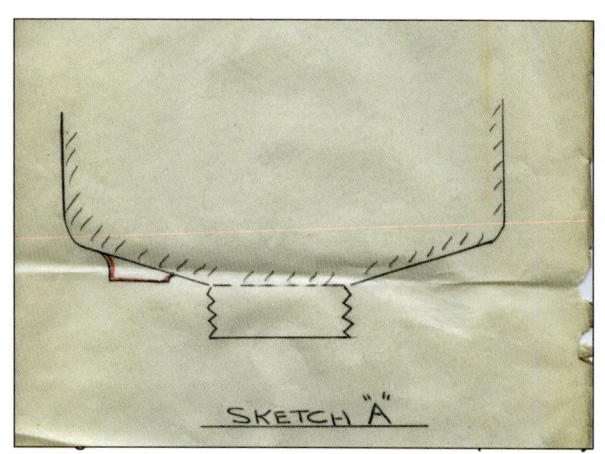

the Committee at the next meeting at Eastleigh.

Gangway. It was considered that the loose canvas fabric of the sample gangway was unsightly when the gangway was closed back against the coach and that the fabric should be corrugated. This would improve the appearance and Mr Turbett will arrange for the gangway to be altered.

Inside Gangway Door. The representatives of the Electrical Department expressed themselves satisfied with this arrangement.

Doors of Switch Cupboard. The representatives of the Electrical Departments agreed to the proposed use of the hinged doors on the switch cupboard in the Driver's compartment on which will be mounted a portion of the door jamb for taking the inside gangway door when open.

It was arranged that the doors should be supplied drilled to allow for the fitting of the door jamb. This work to be done in a manner to be approved by the Electrical Engineer for New Works.

Driver's Compartment. Hand-brake. It has been practice to fit the hand brake in the Driver's compartment. This will not be possible in the design of the cab for the Portsmouth Fast Service Stock, and it was agreed by the representatives of the Electrical Departments and Traffic Departments that there would be no objection to the hand brake being fitted in the Guard's compartment. It was proposed to fit the brake in a position adjacent to the Guard's seat.

Stencil Rack. This will also be required to be fitted in the Guard's compartment. The Electrical Department's representatives saw no objection to this arrangement.

Brake and Electrical Fittings. In view of the restricted space in the Driver's compartment, when the hinged door is in the open position, it is considered desirable that that the compartment should be fitted completely with the brake and electrical equipment for inspection, and that the fittings should be placed as far as possible in a similar position to that of the Eastbourne Stock, and as shown on drawing E.22990. Mr Roberts stressed the importance of placing the driver's equalising reservoir in a position not more than 2 ft. from the driver's valve. The CME Department to keep this before them when erecting the fittings in position in the cab.

Communication between driver and Guard. As it may be necessary in the event of a failure of the power brake to work stock to a convenient station or depot with the Guard operating the hand brake, it was decided that the fixed glass panel in the partition separating the Guard's and Driver's compartments should be replaced by a glass hinged flap, which could be opened to enable audible communication to be made between the Driver and Guard.

Construction of Stock. Mr Turbett stated it was a matter of considerable importance that a definite decision be arrived at as early as possible on the matters outstanding, in view of the limited time remaining for the construction of the stock, which will require to be turned out of Shops by December next.

Next Meeting. It was agreed that the next meeting should take place on Friday 3 April 1936 at 11.30 am. (The document included a list of details required from the Electrical Engineer:

Control switch / Ammeter 6 in. / Speedometer 6 in. / Duplex Gauge 6 in. / Whistle Valve Type E No. 1646 / Air Pressure Reducing Valve / 1 in. Brake Valve isolating cock / Relay and Shunt box / Emergency application valve / Motor Generator starting switch and fuse with blow out coil / Double pole control switch / 10 point control cut out switch / Isolating switch control / Compressor governor / Window wiper engine / Driver's feed valve / Heater / Dead Man's valve.

This was quickly followed by a report from the Traffic Representatives to the Superintendent of Operation just two days later on 28 March 1936.

Express Stock Motor Coaches fitted with Driver's compartments.

At the meeting held at Eastleigh on Thursday, 26 March, at which representatives of the

Departments concerned, namely, Chief Mechanical Engineer, Electrical Engineer for New Works, Electrical Engineer (Running) and Superintendent of Operation, attended, the position of the route indicator on express stock motor coaches provided with a Driver's compartment was considered, and it was the opinion of all concerned that the most satisfactory position for the route indicator was on the offside of gangway on a level with the Driver's lookout window. Three fitments were submitted for inspection, and the one which was considered to be satisfactory projects 5¾" on the offside and 2¼" on the nearside from the face of the coach, which ensures that the indicator will be square with the running lines and gives a clear view at a distance of 15-ft. from the front of the coach at a point 8-ft. outside the running rail. A rough sketch illustrating the points made is shewn below

The gangway when closed projects approximately 9" from the coach face.

The adoption of this arrangement will admit of the use of the standard fittings, including the indicator slides.

No indicator for use above the gangway was provided for inspection, but, if necessary, the question of the suitability of such fitment from a traffic working point of view can be considered at a further meeting at Eastleigh of the representatives of the Departments concerned, which has been tentatively fixed for Thursday of next week.

On the last day of March comes a short note headed from the Chief Mechanical Engineer - so presumably Maunsell although the initials do not necessarily appear to correspond.
Express stock motor coaches fitted with Drivers compartments.

Referring to the meeting held at Eastleigh on Thursday, 26 instant, from a working point of view I see no objection to the route indicator on express stock motor coaches fitted with Drivers compartments being placed on the off side of the gangway on a level with the Driver's lookout window, and the fitment provided for inspection at Eastleigh on the date above mentioned is satisfactory except that I am of opinion it would be an advantage if it can be arranged for the front of it to be in line with the front of the gangway, provided this can be done it without disfiguring the appearance of the coach.

I am advised that no indicator for use above the gangway was available for inspection at the meeting and I shall be glad to know whether you are pursuing this phase of the matter and if the reply is in the affirmative perhaps you will be good

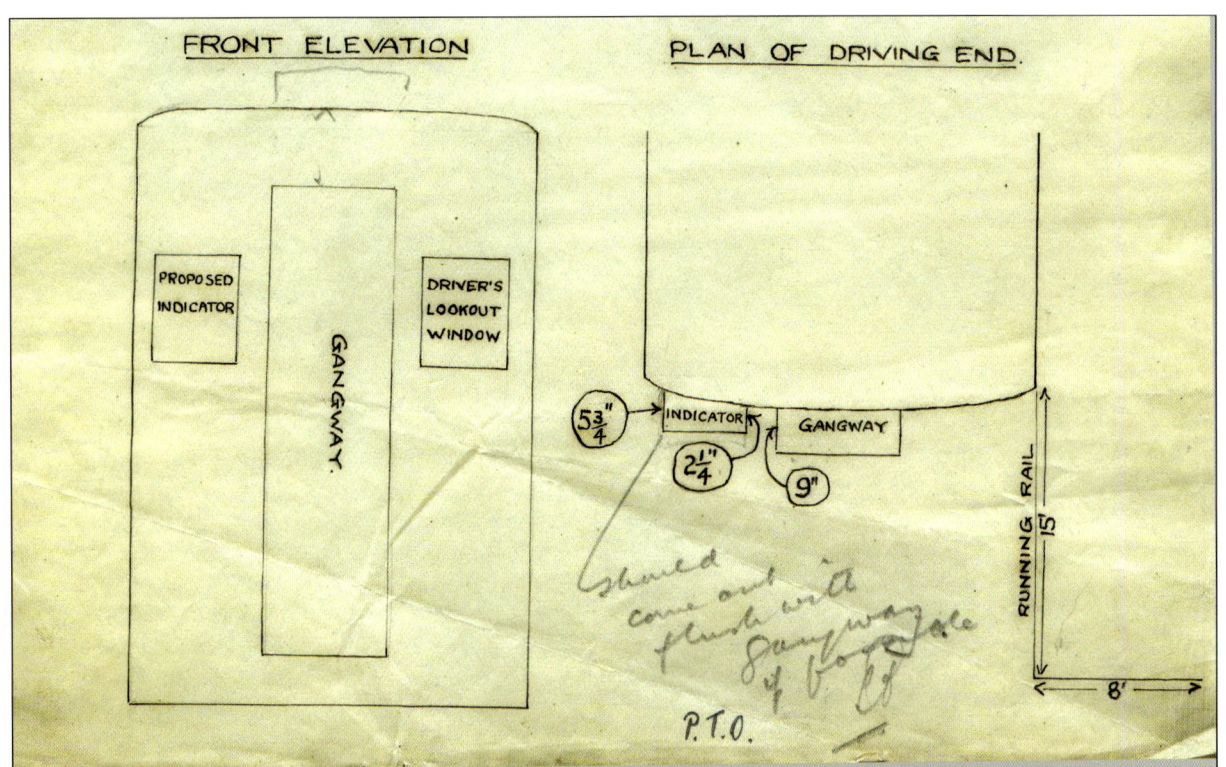

Sketch accompanying the report of 28 March. The original had likely been creased when the papers were last referred to more than 80+ years ago; careful use of the domestic iron has removed most of the damage. The annotation 'PTO' is a puzzle as there was nothing shown on the reverse side.

Left: Sylvan scenery on the Portsmouth direct. As unit 3142 heads a Waterloo to Portsmouth Harbour semi-fast; complete with roof boards.

Opposite bottom: Towards the end of its life, set No 3132 is in green but with the first yellow warning panel added. The stencilled numbering has also given way to a roller blind. *John Wenyon*

enough to let me know when you have evolved a suitable fitment so that it may be inspected.

A further meeting follows, again at Eastleigh and on 3 April 1936.

Present

Mr L Lynes	CME Department
Mr G Key	CME Department
Mr F Munns	CME Department

Mr W C Moore   Electrical Engineer for New Works Department

Mr E G Humfress Electrical Engineer for New Works Department

Mr A E Roberts Electrical Engineer's Department

Subject:- Inspection of model of driving end of the Express Stock for Portsmouth services.

Indicator Box. Mr. Roberts stated that the position of the indicator box was satisfactory so far as department was concerned. If it is advanced from the position shown on the model towards the face of the gangway, it would be inconvenient to the motorman when hanging the slides.

The route indicator box is fitted on the off side of the vehicle, and it was noted that the letters, which are carried in the box, are clearly visible at an angle of 45° when standing on the near side of the vehicle.

The Traffic Department should be asked to say whether this will meet their requirements

Relay Box. The relay box should be placed 10" higher. As a result of this decision, it will be necessary for the control line conduit to be let in into the end framing behind the relay box. The CME Representative said there will be no difficulty in meeting this requirement.

The auxiliary and control conduits from the relay box to the control communication box, to pass over the door head, and down alongside the pillar carrying the hinged gangway door. Mr Key to alter framing to allow the conduits to be 'housed'.

Equalising Reservoir. Mr. Roberts approved its position, which is approximately 4' 0" away from the driver's feed valve. This cancels the decision of the previous meeting.

Electrical Equipment. The Electrical representatives expressed themselves satisfied with the position of the electrical equipment, with the exception of the compressor governor. This is to be placed in a similar position to that on the Eastbourne Stock, and the small observation window to be adjusted accordingly.

Deadman's Valve. Mr Roberts was satisfied with the position of this fitting.

Jumper Connections. The nose end of the model in the vicinity of the jumper connections is steel panelled, and in this respect follows the Express Stock built for the Brighton and Eastbourne Services. Mr Roberts stated that owing to complaints received from the staff working the trains when in the depots, this practice should be discontinued and the Electrical representatives asked the CME to arrange for teak members to be provided for mounting their fittings. This practice will then accord with that which has been adopted for the Suburban Stock.

Lamp to be provided in the section of the Driver's compartment which will be used by the public as a corridor between the units.

The lamp to be on the main cable circuit, and permanently lighted.

Roller Shutters. In Mr Roberts' opinion, the roller shutter fitted to Coach No. 5630 was preferable to the four fold door now fitted to the model. The CME is not satisfied that the shutter displayed is suitable for the purpose and will investigate various patterns of roller shutters before making a decision. It was also suggested that a coach fitted with roller shutters and engine propelling it should be run at high speed to see if the air pressure against the slats would reduce the vibration.[1]

Next and again just three days on, a further and this time somewhat caustic note passed from the CME to the Traffic Manager.

Referring to your letter of the 31st ultimo. The position of the route indicator has been the subject of careful enquiry on two occasions by the Departments concerned.

On the first occasion, viz 26 March when your Mr. Finch was present, it was decided that it would be unsightly if the front of the route indicator frame was built out to be level with the gangway face and it was also considered unnecessary.

A second meeting, which you were aware would take place, was held yesterday, but your representative was not present. I enclose the minute of this meeting, from which you will see the position of the route indicator box was decided upon, subject to any objection you may have to make, and I hope to hear that you are in agreement with the views of the Investigation Committee.

The heat or was it the urgency then appears to go out of the situation as there is now a gap of two weeks before on 18 April the CME again writes; we assume to the Traffic Manager but no addressee is shown on the copy correspondence.

Referring to your letter of the 4th instant, I note your remarks relative to the route indicator frame proposed to be fixed on the off side of the gangway of the new electric coaching stock. I presume however, that the question of providing an indicator at the top of the gangway, as mentioned by your assistant, Mr Turnbull to my Assistant for New Works on the 2nd idem, is receiving attention, and that you will advise me further in regard thereto in due course. If this proposal does not prove to be practicable I think further consideration should be given to bringing the indicator more in line with the front of the gangway.

Almost six weeks later Maunsell appears to have conceded; perhaps he was simply physically tired as he would retire soon after. The following note seeming to bring to a conclusion the whole saga.

From CME to Traffic Manager, 28 May 1936.

Express Stock Motor Coaches fitted with Drivers' Compartments.

Referring to your letter of the 25th inst., and previous correspondence, the question of providing an indicator on the top of the gangway has been investigated., but any possible arrangement of this kind, does not commend itself. Arrangements have, therefore, been made to fix the standard type of route indicator on the off side of the gangway in the position recommended by the representatives of the various departments (including your own) who inspected the model end at Eastleigh. It was then considered unnecessary to bring the indicator forward in line with the front of the gangway and any such arrangement would, in my opinion, be very unsightly.

So all of this over basically just one item. We only wonder if other aspects of the built would receive such scrutiny, in all probability yes!

1 - Four Restriction 1 Maunsell composite coaches, Nos 5630/31/32/33 were fitted with roller shutters about 1934. They were otherwise standard Maunsell compos to SR Diagram 2302. Coach expert Mike King, who has been consulted about this topic, admits he has never seen an image or drawing of the modified vehicles. There is no indication how long the shuttering remained in place, not whether the referred to propelling trial took place. Again according to Mike, 'The 4-COR's had an interesting leading end gangway connector - and most definitely not to Pullman car standard! 'Extrasway' is how I would have described it.' We should conclude by stating the stencil route indicators were replaced in their final years with roller blind numerals but still in the original position.

Variety (or the lack of it) in 1966
Illustrations by Tony Harris

We don't think anyone could really deny there is little variety on today's railways. Until recently if you travelled on most main lines, it would be uncommon to not encounter an HST set – for 2023 read '800' type set. Standardisation may be all well and good but the one size fits all policy is not always the best solution. Perhaps it is a product of passing years that one hankers back to more choice (and from your Editor's perspective he would be prepared to pay a little more for the privilege).

So we go back in time – after all is that not what Southern Times is all about – nearly sixty years to the 1960s when there was variety. Or was there…?

For a little time now we have been privileged to be the repository for a number of photographs taken by Tony Harris and who kindly grouped these into batches / dates / locations (which also does make filing so much easier). We have selected a small group taken by Tony at Southampton on 30 April 1966 but on close examination look carefully and even then there was limited variety…. (Perhaps we are secretly missing all the various shapes and sizes of the Southern moguls, the various Southern 4-6-0s and the almost countless number of classes that once existed from the pre-grouping companies.)

Today's enthusiast may think there is variety in the 21st century. You have my sympathy, you should have seen the 20th century – and I for one wish I had seen more of it!

By April 1966 major inroads had been made into the number of Merchant Navy class engines still serviceable and operating. Here we see No 35030 still carrying its *Elder Dempster Lines* nameplates at the head of a down passenger service and the duty No 383. Fifteen months later to the day, this same engine would have the dubious distinction of hauling the final steam up passenger service into Waterloo on the afternoon of Sunday, 9 July 1967.

Although freight around the Southampton area was basically limited to two types of movement, that from Southampton Docks and oil trains from Fawley, there were still occasional other workings. Here No 34066 *Spitfire* is seen at the head of an up goods about to arrive at Southampton Central.

Normally rostered as a Class 8 Merchant Navy turn, Nine Elms have instead turned out No 34001, formerly carrying the name *Exeter,* for the prestige Bournemouth Belle service seen shortly after leaving Southampton Central en-route to Bournemouth. The time would be soon after 14.00, the engine seemingly in fine fettle with steam to spare. In days past an MN would have almost been guaranteed for this working but as time passed even a Class 5 and of course a Brush Type 4 might be seen. Note the clean third-rail pots and in the background the chimneys of Southampton power station which had its own rail connection to receive coal.

Typical of the run down appearance of steam in 1966 (and indeed in the years before this) was No 73087, reduced to the role of hauling a Romsey bound local leaving the station. Crews found these engines very free running but prone to draughts when running fast.

Another down local, this time for the Bournemouth line behind Class 4 tank No 80085. Behind is what may well once have been a three coach set of either Bulleid or Mk1 stock but as can be seen now, a bit of a mixture.

We conclude the steam views with an interloper from the LMR. Black 5 No 44942 is passing on a Poole to Newcastle service, such trains once the province of a King Arthur or Lord Nelson but those days were past and even a Bulleid or Standard was the choice in 1966.

Finally, the future, well the present anyway. No D6550 with empty tank cars for Fawley and in smart external condition compared with its steam cousins. (The diesels were not always this clean.) The train would have come south via Reading and Basingstoke, its previous route from Didcot and Newbury closed two years earlier. Tank car trains would be a regular sight from Fawley for many years to come, although in more recent times fuel has been moved by underground pipeline.

Treasures from the Bluebell Railway Museum
Tony Hillman

Internal Southern Accident Reports

All readers will be aware of the Official Reports produced by HMSO after a railway accident. What are not so common are the Internal Accident Enquiry Reports following accidents which are not as serious as those needing a full Enquiry.

Sadly, many of these detail accidents to staff, many fatal. Also, minor derailments, collisions and breakaways are covered. The reports often run to many pages, usually including a sketch plan of the area concerned.

Of interest to us, with its Bluebell connections, is a collision at Wadebridge. The Enquiry introduction states:

'At approximately 7.23 am on Monday, 12 August 1957, light engine 34023*, West Country Class *(Blackmoor Vale)*, moved from the Motive Power Depot over No 2 Siding and had reached the "One-Way" hand operated spring points (trailing in the direction of travel) leading from the Motive Power Sidings to the Loop Siding when it came into collision with 0298 Class Tank Engine No 30587 which was propelling three empty wagons along the Loop Siding from the direction of Wadebridge East signal box.

There was considerable damage to the Tank Engine as a result of the collision, but fortunately there were no serious injuries sustained by any staff concerned.'

* *No 34023 is owned by the Bulleid Society and based at the Bluebell Railway.*

The 10-page report records the evidence from those involved, the drivers, firemen, shunters, porter and signalman and a Summary of Evidence written by A. W. Johnston, District Motive Power Superintendent, and W. G. Coward, Assistant to the District Traffic Superintendent, Exeter Central. This runs to nearly three pages and ends with their conclusion where fault lies.

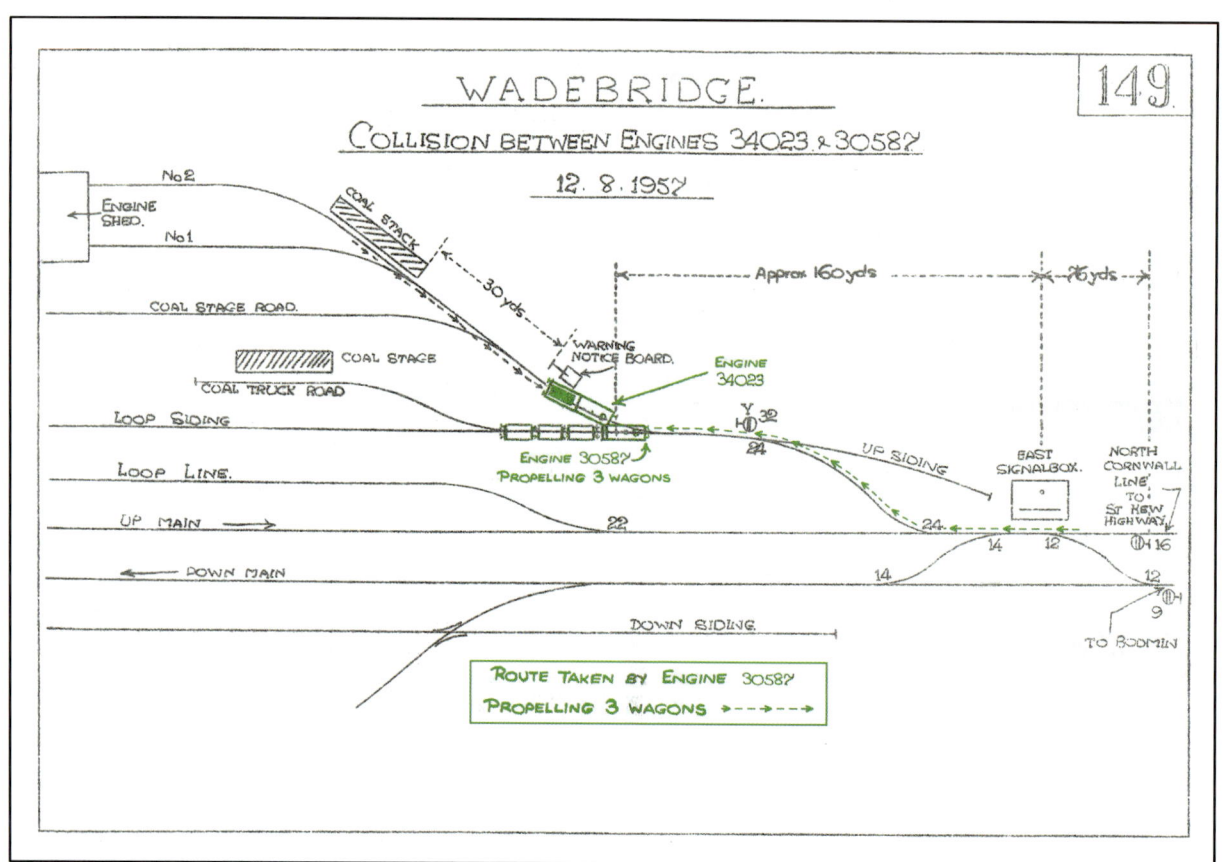

No 30587 after its contretemps. On the blocks at Wadebridge as it was unfit to travel any distance for repair.

It seems everyone involved was at fault in some way. While the fireman on No 34023, who was driving, caused the collision, all the others had failed to correctly apply the Rule Book in some way or another. The fireman on No 34023, and the driver who was standing on the locomotive steps, admitted they could not see ahead because of the amount of steam coming from the cylinder drain cocks. It cleared just in time for the driver to shout a warning, but it was too late to stop a collision. The Shunter and a Porter helping him were riding on No 30587 and did not notice that No 34023 was moving. Even the signalman was at fault in not insisting that before the shunting movement No 30587 ran sufficiently far forward on to the North Cornwall line as required in the General Appendix.

The report concludes with Appendixes.

Injuries are recorded in Appendix B, No 30587's Driver grazed his left leg, one of the Shunters sprained his wrist and the other bruised his right leg.

Appendix C records that the 7.5 am Bodmin Road to Padstow was delayed by 11 minutes and the 8.3 am Launceston to Padstow by 5 minutes. (The 7.5 am time is incorrect, it should be 7.50 am.) The Launceston train is the 1.15 am ex-Waterloo.

Damage to the engines is recorded in Appendix D. No 34023, "Bent engine buffer beam" and "Air flow casing bent on right side". No 30587 suffered far worst with "Leading and trailing axles bent", "Left side spectacle plate and cylinder flange broken", "Side framing and angle iron bent and back draw hook broken".

In his evidence the driver of 34023 states: 'I found that the shunting engine had been badly damaged, but I was able to pull my own engine back clear of it. I then discovered that the shunting engine had not become derailed and I was able to move it slowly up the Loop Siding to clear the Loco exit. I was unable to perform my booked duty as pilot of the 7.46 am Passenger to Padstow, but arrangements were made for me to run light to that point later.'

The Engine Workings and Working Timetable show that No 34023 was working Wadebridge Duty No 609. Officially this leaves Wadebridge Shed at 7.57 am and runs light to Padstow, arriving at 8.06 am. However, presumably to save line occupation, No 34023 was going to leave the Shed in time to pilot the Class O2 on the 7.22 am Bodmin North to Padstow. This arrives at Wadebridge at 7.41 am. Presumably, No 34023 was later able to run light engine to Padstow to haul the 9.35am.

No 34023 will haul the 9.35am, Atlantic Coast Express (ACE), to Waterloo as far as Exeter Central, arriving at 12.12 pm. Two crew changes take place en route. No 34023 then departs for Exmouth Junction shed arriving at 12.24 pm, the end of its day's work. It would seem likely that repairs were then carried out there. The ACE will be taken forward to Waterloo by a Merchant Navy, arriving there at 3:32 pm.

SOUTHERN TIMES

ISSUE 5

Opposite top: No 34023 *Blackmoor Vale* at Wadebridge, in happier times, July 1963. *George Reeve, Irwell Press collection.*

Opposite bottom: No 30587 visiting the Bluebell Railway, Branch Line Weekend in March 2019. *Roger Cruse, Bluebell Railway*

Above: Original items, including the BR smokebox carried by No 30587 and now part of the Bluebell Museum collection.

Next time from the Bluebell Museum Archive: The Southern in WW2

Just for amusement…

Just for amusement, four more stations to identify…?

Clues? Well (in numerical order) try Sussex, Kent, Berkshire and London. (And do bear in mind with No 4, attitudes were different 70 years ago to how we view things today.)

Answers next time.

From the footplate

This time a joyous amount of comments and additions from readers. We start with a letter received from Chris Small in Widnes. (Not sure the Southern ever reached quite as far as Cheshire but Chris, we know, is an emigree from Hampshire.) Chris refers to Issue 2. 'The original Hampshire units (1101-1118) were, as you correctly say, two cars. These were followed by Nos 1119-1123 for East Sussex branches. Nos 1101-1118 were augmented to three cars in 1959 and followed by new built 1123-1126 and 1127-1133 in later years. The problem with three car sets on the Mid Hants has been well documented and it was customary from 1961 to provide two-car sets on the majority of services by the simple expedient of removing the centre car from two random units at Eastleigh depot for a week or two. In May 1964 a reshuffle to provide diesel services on the Steyning line resulted in units 1114-1118 leaving Eastleigh for St Leonards and two car No 1122 coming the other way. No 1121 followed in another exchange in November 1964 and both these were the regular Alton line units until closure. No 1120 was never a regular performer on the line; in fact I have no record of it ever appearing.

'The illustration of No 1124 near Totton is of interest and can definitely be dated between 1 January and 14 June 1967. For two or three years prior to the Bournemouth electrification, certain Saturday and Sunday steam stopping services were scheduled to be worked by three-car DEMUs for which two new route head codes were officially allocated.

03 – Reading or Basingstoke to Bournemouth West

04 - Reading or Basingstoke to Branksome and all stations to Weymouth.

(In reality most trains started from Eastleigh or Southampton.)

'Concerned about poor performance of its main line services on the Bournemouth line prior to the start of electric train operation, the S W Division decided to reduce a further four units from three to two cars and these, together with Nos 1121 and 1122, were used from January 1967 on the Romsey – Southampton and Mid Hants lines. As before, some sets were changed over, but No 1124 remained as two cars throughout this period.'

Now from Richard Newman and again re Issue 2. 'There is a very tiny error in the commentary to the Uckfield station plan shown in the article in *Southern Times* 2. I correct it just in case it crops up again in future. The plan does not show the name of the landowner correctly as Streatfield. This is an ancient family in Kent, centred on Chiddingstone.'

Now re Issue 3 from Rodney Lissenden. Page 30: 30862 is not *Lord Rodney* (I should know the number off by heart) but *Lord Collingwood*. Page 77: the bottom shot is Folkestone to Victoria not as stated, if it had been a down train the 56 would have been in blank red. It is an interesting picture taken off the road bridge by Tonbridge station, the Hastings line curving away to the right and the old steam shed with Class 33 in residence.'

Rodney also points out some potential errors in the shed allocations we have been serialising. (In our defence we should add they are copied as written by the late Alan Elliott.) However for the sake of accuracy the changes are: Battersea; 1062 and 1149 are F1 not E1, 2198 should read 2108 E2. Dover: 1108 is 0-6-0, I think 1189 should read 1198 as no such loco, 1340 down twice, 1445 is B1 4-4-0 not R1. Faversham: 1442 and 1452 B1 4-4-0 not 0-6-0T, 1673 - 1698 should be 0-4-4T. Gillingham: 1446-8 are B1 4-4-0 not R1, 1660 is 0-4-4T not 0-6-0, 1685 is S not C 0-6-0ST (rebuilt). Maidstone West: 1675 R is 0-4-4T, 1699 0-4-4T not 0-6-0T, 1700-1707 are 0-4-4T not 0-6-0, 2273 not E4, should it be D1 0-4-2T?'

Now from Stuart Hicks (ST3). 'Caption on page 77, lower. This image at Tonbridge appears to me to be an UP boat train coming from the Paddock Wood direction (Hastings line to right), not a down train as suggested.'

Now from Denis Calvert. Ref 'From the Footplate' in ST3 and the photo p79 of a 6L DEMU in Tunbridge Wells Central. 'I used (when I was but a very little lad) to live in that town and occasionally to photograph the railway there. That, of course, was in the time that Tunbridge Wells West station still retained much of its former splendour…... . Attached a shot taken in 1966 from above the portal of Grove Hill tunnel. The unit Hastings-bound on the main line is 6L 1016 (as I read it) while, on the right, is the line which diverged to the single-track Grove Tunnel to emerge into the vast expanse of Tunbridge Wells West yard, station and MPD.'

From Alastair Wilson. 'Just finished Southern Times Issue No. 3 - as ever, with great pleasure. I hope this additional information will be of interest.

'Sean Bolam's delightful painting of Droxford had your commentary to the effect that Droxford's greatest moment came when Winston Churchill

Missed out from Issue 4 (we had 'filed' it a bit too carefully) is this view of T14 No 443 on a Bournemouth line service. The location is not confirmed, although it is 'similar' to several other contemporary views taken near St Cross.

and other Allied leaders came to the station to discuss plans for invasion, prior to D-Day.

'The reason for the choice of Droxford was the Supreme Allied Commander was US General Dwight D. Eisenhower (later President Eisenhower, 1953-61). His HQ before D-Day was at Southwick House, which had been requisitioned by the Royal Navy in 1941. The nearest railway station was, in fact, Wickham, also on the Meon Valley line, but it was busier (if any station on the Meon Valley could ever be said to be busy), and what was needed was a place where the temporary stabling of a long train could occur with the minimum of disturbance to other traffic. So Droxford, the next station to the north, was chosen, being about 8 miles as the crow flies from Southwick House.'

From Alan Postlethwaite re ST3.

'The mystery photo on p69 is just south of Shepton Mallet. The main clue is the tall lattice signal post in LSWR style with repeating arms, found commonly on the S&D. Where does another railway cross the S&D? - why, Shepton Mallet, of course. I hereby claim a hat-trick of answers to your mystery photos, having previously identified the northern approach to Streatham in ST1 and then Wimborne in ST2 - an accomplishment surely worthy of mention in your next Footplate section plus a packet of Smarties?' *('Smarties'? – think of the calories….Ed!)*

We have had a long and very pleasant email from Pawel Nowak who noted the farewell tone of Southern Way 57. Perhaps with the passage of some 18 months, now may be the time to summarise the situation as took place. In October 2021 I gave three months' notice to Crecy Publishing that I would cease to work for them from the end of December. I would complete and submit a final issue of Southern Way, No 57, for release in January 2022. Proofs were received in due course and which included a gentle 'farewell' in the submitted introduction for No 57. I was not critical of any organisation or individual, my only comment that I felt now was the time to move on. When No 57 did appear, my complete introduction had been removed. I will admit I was saddened as many readers will therefore have been unaware. (Elsewhere in No 57 I had referred back to this being my final issue but it probably made little sentence without the missing introduction. The rest, as they say, is history…. .)

Pawel writes, 'I was a little worried about the farewell tone of Southern Way No 57, in case it represented a complete retirement or possibly a serious health problem. I am relieved to find you have arrived at "Pastures New" i.e. Transport Treasury, and am most impressed by Issues 1 and 2 and wish you and the team success for the future.

An Arthur King photograph of DEMU 6L set No 1916 at Grove Junction in 1966.

Obviously I won't abandon Southern Way, assuming it keeps going, as I treasure any information about the history and operating practices of the railway on which I started in September 1969 from whatever source. You may be amused that, following various unsuccessful property machinations by our Northern-based property department, I have ended up at the former Hither Green Locomotive Shed, of which the western half survives as a maintenance facility for track machines, as does the SR rail-post mounted clock, which features on countless photographs from the 1950s. My current 1990 office replaced the former 3 eastern side shed roads and the very prominent water tower, but otherwise the actual track layout is largely identical to its steam-age predecessor, apart from the relatively recent removal of the turntable and the lowering of the former coaling stage siding to the surrounding ground level. In my childhood period the place was inhabited by "C", "Q1", "S15" and "W" classes, not to mention the solitary "King Arthur" 4-6-0 allocated to a peak-hour passenger turn from Cannon Street, but it was much later that I actually visited the Marshalling Yard in which I did my early freight training, now retaining a very small freight presence, but mainly a depot for Networkers and other more modern EMUs.

' With regard to the location questions in "Southern Times", I was completely flummoxed by the picture on Page 74 of issue No 1 showing the two berthing sidings with a 4SUB in one and 4EPB in the other. I'm confused by the signals because I can't find in my collection of diagrams or in my memory any pair of sidings with a 3-way junction signal on the approach to them in the manner shown. It must be somewhere within 20 miles of central London judging from the rolling stock and the inter-war housing to the right of the picture, but I still can't place it.

' I think I'm on slightly more solid ground in the case of the picture at the top of Page 75, with a typical 1932 twin-headed junction signal in the picture and a 6PUL unit with headcode "5", denoting London Bridge - Brighton via Quarry Line avoiding Redhill. By coincidence, your issue No2 has a review on Page 63 of the comprehensive signalling history by Chris Durrant of the southern half of the Brighton Main Line. *(To be continued in issue 6.)*